DC Electrical Circuits Wc

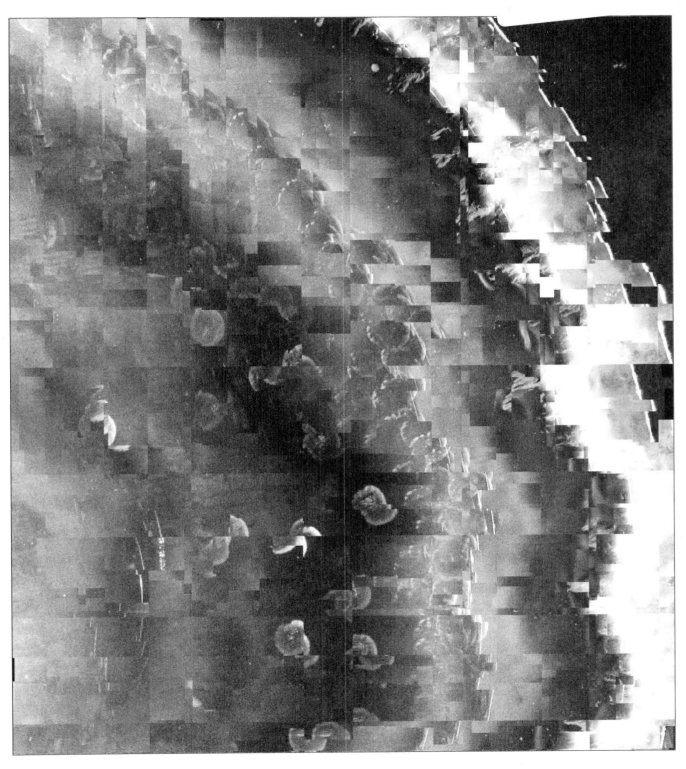

James M. Fiore

DC Electrical Circuits Workbook

by

James M. Fiore

Version 1.2.11, 14 April 2020

This **DC Electrical Circuits Workbook, by James M. Fiore** is copyrighted under the terms of a Creative Commons license:

This work is freely redistributable for non-commercial use, share-alike with attribution

Published by James M. Fiore via dissidents

ISBN13: 978-1796779035

For more information or feedback, contact:
James Fiore, Professor
Electrical Engineering Technology
Mohawk Valley Community College
1101 Sherman Drive
Utica, NY 13501
jfiore@mvcc.edu

For the latest revisions, related titles and links to low cost print versions, go to
www.mvcc.edu/jfiore or www.dissidents.com

Cover art, *Some Thing*, by the author

Introduction

Welcome to the *DC Electrical Circuits Workbook*, an open educational resource (OER). The goal of this workbook is to provide a large number of problems and exercises in the area of DC electrical circuits to supplement or replace the exercises found in textbooks. It is offered free of charge under a Creative Commons non-commercial, share-alike with attribution license. This workbook has been replaced by the new text, *DC Electrical Circuit Analysis*, also OER. The new text features greatly expanded explanations and example material. As such, this workbook will no longer be updated.

The workbook is split into several sections, each with an overview and review of the basic concepts and issues addressed in that section. These are followed by the exercises which are generally divided into four major types: analysis, design, challenge and simulation. Many SPICE-based circuit simulators are available, both free and commercial, that can be used with this workbook. The answers to most odd-numbered exercises can be found in the Appendix. A table of standard resistor sizes is also in the Appendix, which is useful for real-world design problems. Version 1.1 adds material on dependent sources in Section 6 as well as exercises that can make use of supernode/supermesh techniques. If you have any questions regarding this workbook, or are interested in contributing to the project, do not hesitate to contact me.

This workbook is part of a series of OER titles in the areas of electricity, electronics, audio and computer programming. It includes five textbooks covering DC and AC circuit analysis, semiconductor devices, operational amplifiers, and embedded programming using the C language with the Arduino platform. There are seven laboratory manuals; one for each of the aforementioned texts plus computer programming using the Python language, and the science of sound. The most recent versions of all of my OER texts and manuals may be found at my MVCC web site as well as my mirror site: www.dissidents.com

This workbook was created using several free and open software applications including Open Office, Dia, and XnView.

"Insert pithy, droll or historical quote here."

- The Author

Table of Contents

1 Fundamentals 7

Significant digits and resolution; scientific and engineering notation; definitions of charge, current, energy, voltage, power, efficiency and resistance; energy cost and battery life; resistor color code.

2 Series Resistive Circuits 17

Circuits using one or more resistors in series with either voltage sources or a current source.

3 Parallel Resistive Circuits 33

Circuits using two or more resistors in parallel with either a voltage source or current sources.

4 Series-Parallel Resistive Circuits 45

Circuits using multiple resistors in series-parallel with either a single voltage source or current source.

5 Analysis Theorems and Techniques 59

Superposition theorem for multi-source circuits, source conversions, Thévenin's and Norton's theorems, maximum power transfer theorem, delta-Y conversions.

6 Mesh and Nodal Analysis, and Dependent Sources 79

Series-parallel resistor circuits using multiple voltage and/or current sources; dependent sources.

7 Inductors and Capacitors 101

Definitions for inductors and capacitors, initial and steady-state analysis of RLC circuits, and basic RL and RC transient response.

Appendices

A: Standard Component Sizes 118
B: Answers to Selected Numbered Problems 119
C: Answers to Questions Not Asked 128

1 Fundamentals

This section covers:
- Significant digits and resolution.
- Scientific and engineering notation.
- Definitions of charge, current, energy, voltage, power, efficiency and resistance.
- Energy cost and battery life.
- Resistor color code.

1.0 Introduction

Significant Digits and Resolution

A key element of any measurement or derived value is the resulting *resolution*. Resolution refers to the finest change or variation that can be discerned by a measurement system. For digital measurement systems, this is typically the last or lowest level digit displayed. For example, a bathroom scale may show weights in whole pounds. Thus, one pound would be the resolution of the measurement. Even if the scale was otherwise perfectly accurate, we could not be assured of a person's weight to within better than one pound using this scale as there is no way of indicating fractions of a pound.

Related to resolution is a value's number of *significant digits*. Significant digits can be thought of as representing potential percentage accuracy in measurement or computation. Continuing with the bathroom scale example, consider what happens when weighing a 156 pound adult versus a small child of 23 pounds. As the scale only resolves to one pound, that presents us with an uncertainty of one pound out of 156 for the adult, but a much larger uncertainty of one pound out of 23 for the child. The 156 pound measurement has three significant digits (i.e., units, tens and hundreds) while the 23 pound measurement has but two significant digits (units and tens).

In general, leading and trailing zeroes are not considered significant. For example, the value 173.58 has five significant digits while the value 0.00143 has only three significant digits as does 0.000000143. Similarly, if we compute the value 63/3.0, we arrive at 21, with two significant digits. If your calculator shows 21.0 or 21.00, those extra trailing zeroes do not increase accuracy and are not considered significant. An exception to this rule is when measuring values in the laboratory. If a high resolution voltmeter indicates a value of, say, 120.0 millivolts, those last two zeroes are considered significant in that they reflect the resolution of the measurement (i.e, the meter is capable of reading down to tenths of millivolts).

When performing calculations, the results will generally be no more accurate than the accuracy of the initial measurements. Consequently, it is senseless to divide two measured values obtained with three significant digits and report the result with ten significant digits, even if that's what shows up on the calculator. For these sorts of calculations, you can't expect the result to be any better than the "weakest link" in terms of resolution and resulting significant digits.

Scientific and Engineering Notation

Scientific and engineering notations are ways to express numbers without a lot of trailing or leading zeroes. They also simplify calculations. The idea is to represent the value in two parts: a precision portion, or *mantissa*; and the magnitude, a power of ten called the *exponent*. Thus, 360 can be written as 3.6 times 100, or $3.6 \cdot 10^2$, where 3.6 is the mantissa and 2 is the exponent. Similarly, 0.00275 can be written as the value 2.75 times 0.001, or $2.75 \cdot 10^{-3}$. As the base of the exponent is always 10, a more compact form replaces "· 10" with "E". Thus, these two values can be written as 3.6E2 and 2.75E-3. When adding or subtracting values in this form, the first step is to make sure that all of the values have the same exponent. Then, the precision portions are simply added together. Thus, 3.6E2 + 1.1E2 is 4.7E2. Further, 3.6E2 + 5E1 is converted as 3.6E2 + 0.5E2 yielding 4.1E2. Alternately, it can be converted as 36E1 + 5E1 yielding 41E1 (the same answer, 410 in ordinary form). Where this notation is particularly useful is when multiplying or dividing. For multiplying, multiply the mantissas and add the exponents. For dividing, divide the mantissas and subtract the exponents. For example, multiply 20000 by 360000. This is equivalent to 2E4 times 3.6E5. The result is 7.2E9 (i.e., 7200000000). Similarly, dividing the value 0.006 by 50000 yields 6E−3 divided by 5E4, or 1.2E−7 (or in ordinary form 0.00000012). Notice the cumbersome and error-prone quantities of trailing and leading zeroes in these examples when using ordinary form.

Engineering notation is the same as scientific notation with the caveat that the exponent must by a multiple of 3. Thus, 390000 would be written as either 390E3 or possibly as 0.39E6. Each multiple of 3 has a name and abbreviating letter to make the written representation even more compact. The values and names are:

Exponent	Name	Abbreviation	Exponent	Name	Abbreviation
12	Tera	T	−3	milli	m (note lower case)
9	Giga	G	−6	micro	μ (Greek letter *mu*)
6	Mega	M	−9	nano	n
3	kilo	k	−12	pico	p

Thus, 2000 volts is 2 kilovolts (2 kV) and 0.005 amps is 5 milliamps (5 mA).

Charge, Current, Energy, Voltage, Power, Efficiency and Resistance

Charge is an attractive force. It is denoted by the letter Q and has units of coulombs. Electrons are negatively charged and protons are positively charged. All electrons and protons exhibit the same magnitude of charge, roughly 1.602E−19 coulombs. Thus, one coulomb is equivalent to the charge exhibited by approximately 1/1.602E−19, or 6.242E18 electrons. Further, opposite charges attract while like charges repel, similar to the poles of a magnet.

Current is the rate of charge movement. It is denoted by the letter I and has units of amps (or amperes). One amp of current is defined as one coulomb per second. That is, one amp can be visualized as approximately 6.242E18 electrons passing through a point in a period of one second. Thus, $I = Q/t$. If three coulombs pass through a wire in one-half of a second, this is equivalent to six amps of current.

Energy is defined as the ability to do work. It is denoted by the letter W. The basic unit is the joule although other units are sometimes used (for example, the calorie or the kilowatt-hour, KWH).

Voltage refers to the amount of work (energy) required to move a charge from one point to another. It is denoted by the letter V (although voltage sources often use E, short for *electromotive force* or *EMF*). The basic unit is the volt. One volt is defined as one joule per coulomb, thus $V = W/Q$. If 100 joules of energy are used to transfer 20 coulombs of charge, this is equivalent to five volts.

Power is the rate of energy usage. It is denoted by the letter P and has units of watts, although other units are sometimes used such as the horsepower (1 horsepower = 746 watts). One watt is defined as one joule per second, or $P = W/t$. Thus, if a device consumes 100 joules in 0.1 seconds, the power is 1000 watts or roughly 1.34 horsepower. Power can also be found by multiplying voltage by current. As $I = Q/t$ and $V = W/Q$, then the product $IV = Q/t \cdot W/Q$, or W/t. Thus, $P = IV$. This is known as *power law*. For example, if a 9 volt battery delivers a current of 0.1 amp, the equivalent power is 0.9 watts.

Efficiency is the ratio of useful output power to applied power expressed as a percentage. It is denoted by the Greek letter η (eta) and is always less than 100%. $\eta = P_{out}/P_{in}$. Thus, if a device draws 200 watts and produces 120 watts of useful output, its efficiency is 60%, implying that 40% (80 watts) is wasted.

Resistance is a measure of how difficult it is to establish current flow (i.e., resistance to current flow). It is denoted by the letter R and has units of ohms. Ohms are denoted by the capital Greek letter omega, Ω. The reciprocal of resistance is called *conductance*, G, and has units of siemens. $R = 1/G$ and $G = 1/R$. Resistance is a function of the material the current is passing through along with its shape, see Figure 1A immediately following, where the arrow shows the direction of current flow.

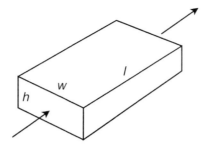

Figure 1A

$R = \rho L/A$, where ρ (rho) is the resistivity of the material, L is the length and A and the cross sectional area (i.e., the area of the face, or height times width). Resistivity is often specified in ohm-centimeters with the length and area similarly specified in centimeters and square centimeters, respectively.

Resistance can also be defined in terms voltage and current. $V = I R$. This is called *Ohm's law* and is one of the most important laws governing practical electrical circuit behavior. Note that Ohm's law and power law may be combined to yield $P = I^2 R$ and $P = V^2/R$.

Energy Cost and Battery Life

Knowing the voltage and current characteristics of a given device allows us to determine its power rating and energy consumption. Based on the per unit cost of energy, the cost to operate the device can be determined. Although most people refer to their local electricity supplier as "the power company", we do not buy "power", per se. Rather, we are billed for energy. Although it would be possible to determine the cost per joule (or more practically, per kilojoule), the supplier normally bills based on kilowatt-hours, KWH. This is because most home appliances are rated in terms of power consumption in watts. Multiplying the power consumption by the length of time the device is used yields a KWH value. For example, a 1500 watt toaster oven used for 30 minutes (.5 hours) yields 750 watt-hours, or 0.75 KWH. If the utility charges 10 cents per KWH, then the cost to run the appliance is 7.5 cents. If it is used for a full hour then it costs 15 cents, and so on.

A battery is a device used to store electrical energy. Ideally, it presents a constant voltage, its current varying according to the resistance of the device attached to it via Ohm's law. Eventually, the energy stored in the battery will be exhausted and its voltage will drop to zero. The storage capacity of a battery is measured in amp-hours, Ah (or milliamp-hours, mAh, for smaller batteries). Thus, a 10 Ah battery can deliver 10 amps of current for an hour. Alternately, it could deliver 20 amps of current for a half hour, 0.2 amps of current for 50 hours, or some other product of time and current that yields 10 Ah. Practically speaking, the efficiency of the battery drops off at very high currents and the lifespan will be somewhat shorter than predicted.

Resistor Color Code

Resistors are devices used to control the currents and voltages in a circuit. They are available in standardized ohmic values and at standardized power ratings (see Appendix A). Along with their resistance value, resistors also have a specified tolerance. This specifies an allowable range of variation of the stated value. For example, a 220 ohm resistor may have a tolerance of 10%. This means that the actual value of any given specimen from a box of these resistors may be off of the nameplate or nominal value by 10% or 22 ohms. Thus, any particular resistor might be as high as 242 ohms or as low as 198 ohms.

While high precision resistors often have their nominal value printed directly on them, general purpose resistors use a color code to denote their value and tolerance. Typically, this will involve four color stripes: two for the precision/mantissa, one for the power of ten, and the fourth to indicate the tolerance. Refer to Figure 1B, immediately following.

Figure 1B

The first two bands, here yellow and violet, indicate the precision or leading digits. The third band, here orange, indicates the power of ten or "number of zeroes" to add. The fourth band, silver in this example, indicates the tolerance. Note that the fourth band is spaced away from the other three to avoid accidentally reversing the order.

The tolerance colors are as follows. A silver tolerance band indicates ±10% while gold indicates ±5%. If the fourth band is omitted, this indicates a tolerance of ±20%. Silver and gold are not used for the precision bands. If they are used for the multiplier band then gold means "divide by 10" and silver means "divide by 100".

The main colors are 0=black, 1=brown, 2=red, 3=orange, 4=yellow, 5=green, 6=blue, 7=violet, 8=gray, 9=white. In the example above, this translates to 4, 7 and 3. The value is "47 with 3 zeroes", or 47000 ohms. The silver fourth band indicates 10% tolerance. Thus, the resistor pictured previously is 47 k ohms with ±10% variation around the nominal value being acceptable. The tolerance yields ±4.7 k ohms, so the acceptable range is from 42.3 k ohms to 51.7 k ohms.

1.1 Exercises

Analysis

1. Round the following to four significant digits: a) 14.5423 b) 30056 c) 76.90032 d) 0.00084754
2. Round the following to three significant digits: a) 354.005 b) 9100.46 c) 1.0054 d) 0.000052753
3. Convert the following to scientific notation: a) 23.61 b) 12000 c) 7632 d) 0.00509
4. Convert the following to scientific notation: a) 4253 b) 640000000 c) 2.03 d) 0.00000658
5. Convert the following to engineering notation: a) 12000 b) 470 c) 6.5 d) 0.00198
6. Convert the following to engineering notation: a) 3500 b) 17.9 c) 5601000 d) 0.0000355
7. What is the charge in coulombs of a million million (10^{12}) electrons?
8. What is the charge in coulombs of 10^{15} electrons?
9. How many electrons would be needed for a charge of 20 coulombs?
10. How many electrons would be needed for a charge of 1 microcoulomb?
11. If a charge of 2 coulombs passes through a wire in 5 seconds, what is the current?
12. If a charge of 300 millicoulombs passes through a wire in 0.1 seconds, what is the current?
13. How much charge must be transferred in 0.1 seconds in order to achieve a current of 5 amps?
14. How much charge must be transferred in 20 seconds in order to achieve a current of 10 microamps?
15. Determine the resulting voltage if it takes 2 joules to move 10 coulombs of charge.
16. Determine the resulting voltage if it takes 15 joules to move 0.5 coulombs of charge.
17. How much energy is required to create a 10 volt potential difference with a 2 coulomb charge?
18. How much energy is required to create a 50 millivolt potential difference with a 0.1 coulomb charge?
19. What is the wattage equivalent of two horsepower (2 hp)?
20. What is the horsepower equivalent of 1000 watts?
21. If a device draws 2 amps of current from a 12 volt battery, determine the power delivered.
22. If a device draws 10 milliamps of current from a 1.5 volt battery, determine the power delivered.
23. A 2 hp motor draws 1800 watts from its source. Determine its efficiency.
24. A 5 hp motor draws 4.5 kw from its source. Determine its efficiency.

25. An audio power amplifier is rated for 500 watts of maximum output at an efficiency of 80%. Determine the amount of wasted power.

26. A compressor draws 10 amps of current from a 120 volt source. Its rated output is 1 hp. Determine the efficiency.

27. An application requires a battery to deliver 15 mA for at least 200 hours. Determine the required amp-hour rating.

28. Determine the required rating for a battery to deliver 0.8 A for at least 30 hours.

29. A certain 12 volt battery has a rating of 6 Ah. Determine the expected battery life using a 5 mA draw.

30. A certain AA battery has a rating of 800 mAh. Determine the expected battery life using a 20 mA draw.

31. Assume that a certain piece of material has a resistance of 80 ohms. Determine the new resistance if the length of the piece is doubled and no other parameters are changed.

32. Assume that a certain piece of material has a resistance of 2 k ohms. Determine the new resistance if the width and height of the piece are doubled and no other parameters are changed.

33. Assume that a certain piece of material has a resistance of 4 ohms. Determine the new resistance if the resistivity is doubled and no other parameters are changed.

34. Assume that a certain piece of material has a resistance of 10 k ohms. Determine the new resistance if the length, width and height of the piece are all halved.

35. A certain material has a resistivity of 100 ohm-centimeters. Determine the resistance of a piece that is 1 cm wide, 0.5 cm high and 6 cm long.

36. A certain material has a resistivity of 2000 ohm-centimeters. Determine the resistance of a piece that is 2 mm wide, 4 mm high and 10 mm long.

37. A 40 ohm resistor has dimension of 0.4 cm wide by 0.2 cm high by 1 cm long. Determine the resistivity in ohm-centimeters.

38. A 5000 ohm resistor has dimension of 5 mm wide by 3 mm high by 6 mm long. Determine the resistivity in ohm-centimeters.

39. A resistor with the color code yellow-violet-red-silver has a measured value of 4806 ohms. Is this resistor within tolerance? As a percentage, how far is it from the nominal value?

40. A resistor with the color code orange-orange-yellow-gold has a measured value of 33.9 k ohms. Is this resistor within tolerance? As a percentage, how far is it from the nominal value?

41. A resistor with the color code brown-black-orange-silver has a measured value of 9980 ohms. Is this resistor within tolerance? As a percentage, how far is it from the nominal value?

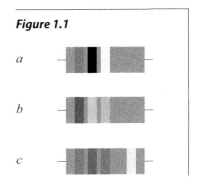

Figure 1.1

42. A resistor with the color code green-blue-black-gold has a measured value of 50 ohms. Is this resistor within tolerance? As a percentage, how far is it from the nominal value?

43. Determine the value of the resistors pictured in Figure 1.1 (left-to-right: red-black-yellow, blue-gray-orange, red-violet-red-silver).

44. Determine the value of the resistors pictured in Figure 1.2 (left-to-right: red-red-orange-gold, brown-black-red-gold, yellow-violet-prange-silver).

45. Determine the value of the resistors pictured in Figure 1.3 (left-to-right: orange-orange-green-silver, white-brown-black-silver, orange-white-brown-gold).

46. Determine the value of the resistors pictured in Figure 1.4 (left-to-right: silver-orange-violet-yellow, green-blue-yellow-gold, gold-black-orange-yellow).

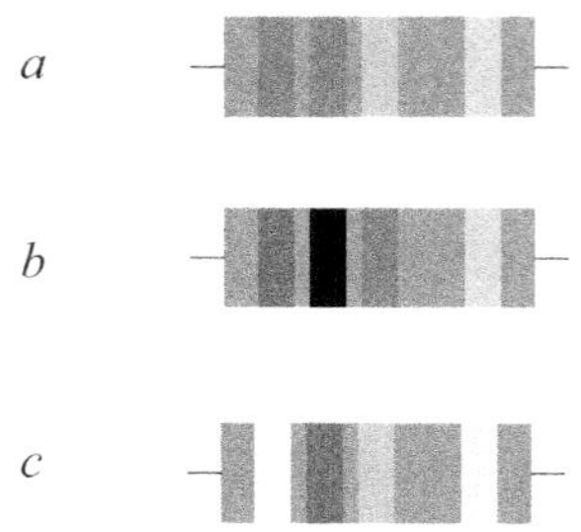

47. Determine the maximum and minimum allowed values of the resistors pictured in Figure 1-1.

48. Determine the maximum and minimum allowed values of the resistors pictured in Figure 1-2.

49. Determine the maximum and minimum allowed values of the resistors pictured in Figure 1-3.

50. Determine the maximum and minimum allowed values of the resistors pictured in Figure 1-4.

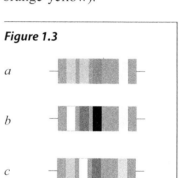

Figure 1.2

a

b

c

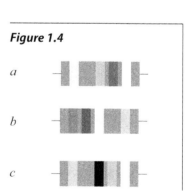

Figure 1.3

a

b

c

Design

Figure 1.4

a

51. Determine the resistor color code for the following ohmic values using 10% tolerance: a) 56 Ω b) 33 kΩ c) 470 kΩ d) 1.2 kΩ e) 750 Ω

b

52. Determine the resistor color code for the following ohmic values using 5% tolerance: a) 47 Ω b) 22 kΩ c) 390 kΩ d) 2.2 kΩ e) 560 Ω

c

Challenge

53. A radio transmitter is rated for 100 watts of maximum output at an efficiency of 90%. If it is fed from a 120 volt source, determine the current draw.

54. A certain 75 watt incandescent bulb produces 71 watts worth of heat and the remainder in the form of light. Determine its efficiency as a lighting device and its efficiency as a heating device.

55. Assume it takes an 1800 watt toaster 3 minutes to toast a bagel "just right". If you toast a bagel every morning for a year and electricity costs 15 cents/KWH, how much will you have spent in that year to toast bagels?

56. Assume that you can buy standard 60 watt incandescent light bulbs for 50 cents each and that each has an expected life span of 1000 hours. In comparison, you can buy an LED light bulb that produces the same amount of light but only consumes 7 watts. The LED bulbs cost $5.50 each and have an expected life of 20,000 hours. Assuming electricity costs 14 cents/KWH, determine the total cost of running incandescent lights versus LEDs for 40,000 hours.

57. Given a 1.5 volt battery with a 500 mAh rating, how much current can it produce continuously for 25 hours?.

58. Given a 9 volt battery with a 100 mAh rating, determine the total energy storage in joules.

Notes

2 Series Resistive Circuits

This section covers:
- Circuits using one or more resistors in series with either voltage sources or a current source.

2.0 Introduction

A series circuit is characterized by a single loop or path for current flow. Consequently, *the current is the same everywhere in a series circuit*. As resistors in series add, total resistance may be found by summing the individual resistors. Multiple voltage sources in series may also be added, however, polarities must be considered as opposing voltages partially cancel each other (i.e., adding a negative). In contrast, differing current sources are not placed in series as they would each attempt to establish a different series current, a practical impossibility.

Along with Ohm's law, the key law governing series circuits is Kirchhoff's voltage law, or KVL. This states that the sum of voltage rises and voltage drops around a series loop must equal zero (the rises and drops having opposite polarities). Alternately, it may be reworded as the sum of voltage rises around a series loop must equal the sum of voltage drops. As a pseudo formula: $\Sigma V\uparrow = \Sigma V\downarrow$

There are multiple techniques for solving series circuits. If all voltage source and resistor values are given, the circulating current can be found by dividing the equivalent voltage by the sum of the resistances. Once the current is found, Ohm's law can be used to find the voltage drops across individual resistors. At that point, power law can be used to find the power dissipation in each resistor or the power developed by the source(s). Alternately, the voltage divider rule may be used to find the voltage drops across the resistor(s) of interest. This rule exploits the fact that voltage drops in a series loop will be directly proportional to the size of the resistances. Thus, the voltage across any resistor must equal the net supplied voltage times the ratio of the resistor of interest to the total resistance, $V_A = E \cdot R_A/R_{TOTAL}$.

If the circuit uses a current source instead of a voltage source, then the circulating current is known and the voltage drop across any resistor may be determined directly using Ohm's law.

If the problem concerns determining resistance values, the basic idea will be to use these rules in reverse. For example, if a resistor value is needed to set a specific current, the total required resistance can be determined from this current and the given voltage supply. The values of the other series resistors can then be subtracted from the total, yielding the required resistor value. Similarly, if the voltages across two resistors are known, as long as one of the resistance values is known, the other resistance can be determined using either the voltage divider rule or Ohm's law.

A key item of importance when analyzing series circuits, or indeed any electrical circuit, is noting the polarity of the voltages. Voltage sources are easy as their polarity is fixed (positive is always at the "long bar" and negative at the "short bar"). The polarity of any resistor will depend on the direction of current flow. Where conventional current enters a resistor, we denote this with a plus sign, and where the current leaves, a minus sign. Moving from plus to minus is referred to as a voltage drop. That is, we are moving from a more positive or higher potential to a less positive or lower potential, thus "dropping" voltage. Similarly, traversing from minus to plus is called a voltage rise.

Along with determining the voltage across single resistors, we are often interested in determining the voltage between points in a circuit. These may span several components. It is imperative that we know the polarities of the individual voltages in order to determine the voltage between any two circuit points.

The foregoing is illustrated using the series circuit schematic of Figure 2A.

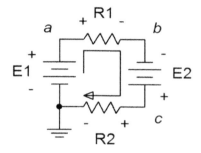

Figure 2A

Here, the two voltage sources, E1 and E2, aid each other, producing a net voltage of E1+E2. Their polarities are fixed. The current direction will be as drawn because conventional current flows out of the positive terminal of a voltage source. The value of this current will be (E1+E2)/(R1+R2) via Ohm's law. Knowing the direction of current, the polarities of the voltage drops across the two resistors may be found. The point where the current enters the resistor is positive, and negative where it exits. Once these are labeled, either Ohm's law or the voltage divider rule can be used to determine the voltages developed across the two resistors. Note that KVL states that the sum of these two voltages must equal the net supplied voltage, or E1+E2 in this case.

To determine the voltage from point *a* to point *c*, or V_{ac}, we start at point *a* and sum the voltage drops and rises until we get to point *c*. Here, the voltage across R1 shows up as positive (+ to −) and the voltage across E2 shows up as negative (− to +). The result is the voltage across R1 minus the voltage across E2. Depending on the specific values, this may wind up being either a positive or negative value. If it's positive, this signifies that point *a* is at a higher potential than is point *c*. Conversely, if it's negative, this signifies that point *a* is at a lower potential than is point *c*. If we reverse the order, starting at point *c* and moving to point *a*, or V_{ca}, we will wind up with the same voltage magnitude but the sign will flip. In the laboratory, the first point letter (the *a* in V_{ac}) is where the red lead of the voltmeter is connected and the second letter is where the black lead is connected. Often, a single connection point is used, as in V_a. In this case, the second letter is assumed to be ground. Thus, V_a is the voltage from point *a* to ground.

To illustrate the importance of voltage source polarity, consider the series circuit shown in Figure 2B.

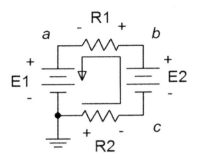

Figure 2B

This circuit is identical to the previous circuit with the exception that the polarity of source E2 has been flipped. This can have a drastic change on the resulting current and voltage drops. For example, if E2 is larger than E1, the net supplied voltage will be E2 − E1 and the direction of conventional current will be as drawn. Note that this is opposite to the prior circuit. Because both the direction and the magnitude of the current have changed, the voltage drops and polarities of R1 and R2 will change. Consequently, any point-to-point voltage, such as V_{ac}, will also change.

2.1 Exercises

Analysis

1. For the circuit of Figure 2.1, determine the circulating current.

Figure 2.1

2. For the circuit of Figure 2.2, determine the circulating current.

Figure 2.2

3. For the circuit of Figure 2.1, determine the power dissipated in the resistor.

4. For the circuit of Figure 2.2, determine the power dissipated in the resistor.

5. Determine the voltage at the open terminals of Figure 2.3.

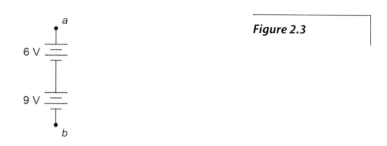

Figure 2.3

6. Determine the voltage at the open terminals of Figure 2.4.

Figure 2.4

7. Determine the voltage at the open terminals of Figure 2.5.

Figure 2.5

8. Determine the equivalent resistance of circuit shown in Figure 2.6.

Figure 2.6

9. Determine the equivalent resistance of circuit shown in Figure 2.7.

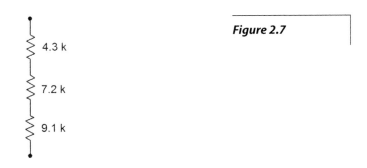

Figure 2.7

10. Determine the equivalent resistance of circuit shown in Figure 2.8.

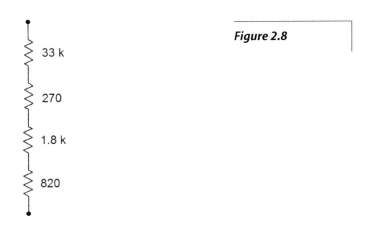

Figure 2.8

11. For the circuit of Figure 2.9, determine the circulating current.

Figure 2.9

12. For the circuit of Figure 2.9, determine the voltages across each resistor and find V_{ab}.

13. Given the circuit of Figure 2.9, determine the power dissipated by each resistor and the power delivered by the source.

14. For the circuit of Figure 2.10, determine the circulating current.

Figure 2.10

15. Given the circuit of Figure 2.10, determine the voltages across each resistor and find V_{ba}.

16. For the circuit of Figure 2.10, determine the power dissipated by each resistor and the power delivered by the source.

17. For the circuit of Figure 2.11, determine the circulating current.

Figure 2.11

18. For the circuit of Figure 2.11, determine the voltages across each resistor and find V_a.

19. For the circuit of Figure 2.12, determine the circulating current and indicate all voltage polarities.

Figure 2.12

20. Given the circuit of Figure 2.12, determine the voltages across each resistor and find V_b, V_{bc}, and V_{ca}.

21. For the circuit of Figure 2.12, determine the power delivered by the source.

22. For the circuit of Figure 2.13, determine the circulating current and indicate all voltage polarities.

Figure 2.13

23. Given the circuit of Figure 2.13, determine the voltages across each resistor and find V_c, V_{ac}, and V_a.

24. For the circuit of Figure 2.13, determine the power dissipated by the 10 Ω resistor.

25. For the circuit of Figure 2.14, determine the circulating current and indicate all voltage polarities.

Figure 2.14

26. For the circuit of Figure 2.14, determine the voltages across each resistor and find V_b, V_c, and V_{ca}.

27. Referring to the circuit of 2.15, determine the voltages across each resistor and find V_b, V_c, and V_{ac}.

Figure 2.15

28. Referring to the circuit of Figure 2.15, determine the circulating current and indicate all voltage polarities.

29. Given the circuit of 2.16, determine the voltages across each resistor and find V_b, V_c, and V_{ac}.

Figure 2.16

30. Referring to the circuit of Figure 2.16, determine the circulating current and indicate all voltage polarities.

31. Given the circuit of 2.17, determine the circulating current and indicate all voltage polarities.

Figure 2.17

32. Referring to the circuit of Figure 2.17, determine the voltages across each resistor and find V_b, V_c, and V_{ac}.

33. Given the circuit of 2.18, determine the circulating current and indicate all voltage polarities.

Figure 2.18

34. Referring to the circuit of Figure 2.18, determine the voltages across each resistor and find V_b, V_c, and V_{ac}.

35. Using the voltage divider rule, determine the voltages V_b, V_c and V_{ac} for the circuit shown in Figure 2.19.

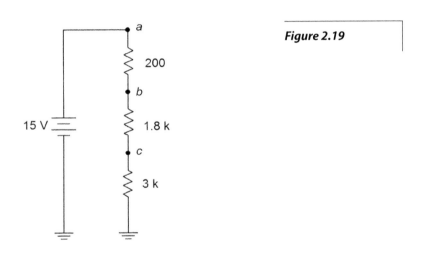

Figure 2.19

36. Using the voltage divider rule, determine the voltages V_b, V_c and V_{bd} for the circuit shown in Figure 2.20.

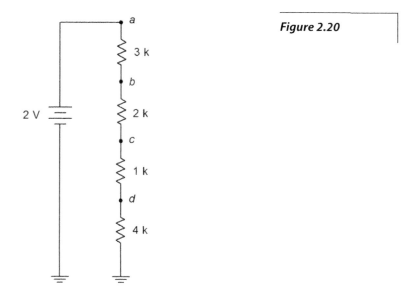

Figure 2.20

37. For the circuit of Figure 2.20, determine V_b if the 4 kΩ resistor is accidentally shorted. How does this compare to the original circuit?

38. For the circuit of Figure 2.20, determine V_b if the 4 kΩ resistor is accidentally opened. How does this compare to the original circuit?

39. Given the circuit shown in Figure 2.21, find the voltage drop across the resistor.

Figure 2.21

40. Given the circuit shown in Figure 2.22, find the voltage drops across the resistor.

Figure 2.22

41. Find the voltage drops across the resistors in the circuit of Figure 2.23.

Figure 2.23

42. Find the voltage drops across the resistors in the circuit of Figure 2.24.

Figure 2.24

43. Find the voltage drops across the resistors in the circuit of Figure 2.25.

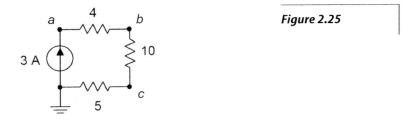

Figure 2.25

44. Find the voltage drops across the resistors in the circuit of Figure 2.26.

Figure 2.26

45. The circuit of Figure 2.27 uses a linear taper potentiometer. Determine V_b when the wiper arm is at position a, position b, and at the halfway point.

Figure 2.27

46. What is the maximum current flowing through the potentiometer of Figure 2.27? At what position(s) does this occur?

Design

47. Redesign the circuit of Figure 2.1 using a new resitor such that the current from the 12 volt battery is 0.1 A.

48. Redesign the circuit of Figure 2.2 using a new resistor such that the current from the 9 volt battery is 2 mA.

49. For the circuit of Figure 2.6, find the value of a series voltage source that would generate 1 mA of current if it was connected across the terminals.

50. For the circuit of Figure 2.8, find the value of a series voltage source that would generate 1 mA of current if it was connected across the terminals.

51. Determine values for the resistors in Figure 2.28 such that R1 is four times the size of R2 and R2 is three times the size of R3, with the total resistance equaling 8 kΩ.

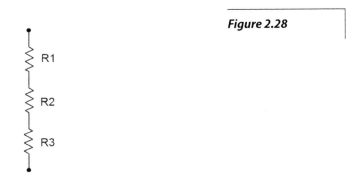

Figure 2.28

52. For the circuit shown in Figure 2.29, determine values for R1 and R2 such that V_{ab} is 6 volts if E is a 9 volt battery and the total current draw is 20 mA.

Figure 2.29

53. Consider the circuit shown in Figure 2.30. If all resistors have the same value, determine that value if E, a 24 volt source, generates a total power of 10 watts.

Figure 2.30

54. For the circuit shown in Figure 2.31, determine values for R1 and R2 such that V_{ab} is 6 volts if I is a 2 mA source and the total voltage drop is 24 volts.

Figure 2.31

55. Consider the circuit of Figure 2.20. Is it possible to add a fifth resistor such that the circulating current is 0.1 mA? If so, what is that resistor value?

56. Consider the circuit of Figure 2.20. Is it possible to add a fifth resistor such that the circulating current is 2 mA? If so, what is that resistor value?

Challenge

57. Assume that two AA cells, E1 and E2, rated at 900 mAh each are used to drive a 2 watt lamp as shown in Figure 2.32. Determine the expected life of the batteries.

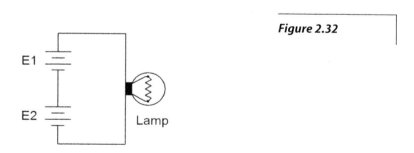

Figure 2.32

58. Given the circuit of Figure 2.33, determine the required values of E, R1, R2 and R3 if there is one volt across R3, the total current draw is 10 mA, the voltage across R1 is twice the size of voltage across R2 and the power dissipation in R2 is 100 mW.

Figure 2.33

59. Given the circuit of Figure 2.33, determine the required source voltage if R1 is 1 kΩ, R2 is 1 kΩ, the power dissipation in R1 is 4 mW and the power dissipation in R3 is 2 mW.

60. Given the circuit of Figure 2.34, determine V_c, V_{db} and V_{ce}.

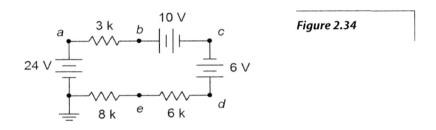

Figure 2.34

61. Given the circuit of Figure 2.34, determine V_{ac}, V_{eb} and V_d.

62. Refer to the circuit of Figure 2.20. Assuming each resistor has a 10% tolerance, determine the maximum and minimum values for V_c.

Simulation

63. Simulate the solution of design problem 43 and determine if the values produce the required results.

64. Perform a DC simulation on the circuit of Figure 2.13 and find all of the node voltages along with the circulating current.

65. Perform a DC simulation on the circuit of Figure 2.14 and find all of the node voltages along with the circulating current.

66. Perform a DC simulation on the circuit of Figure 2.26 and find all of the node voltages.

67. Perform a DC simulation on the circuit of Problem 34 and find all of the node voltages.

68. Perform a DC simulation on the circuit of Problem 36 and find all of the node voltages.

69. Simulate the solution of Challenge problem 58 and determine if the values produce the required results.

70. Simulate the circuit of Figure 2.34 (Challenge problems 60 and 61) and determine if the node voltages produced match the expected results.

71. Perform a Monte Carlo simulation on the circuit of Figure 2.19. Set each resistor to 5% tolerance and run at least ten variations for V_b to determine a typical spread of values.

3 Parallel Resistive Circuits

This section covers:
- Circuits using two or more resistors in parallel with either a voltage source or current sources.

3.0 Introduction

Parallel circuits are in many ways the complement of series circuits. The most notable characteristic of a parallel circuit is that all components see the same voltage. Consequently, parallel circuits have only two nodes. Currents divide among the resistors in proportion to their conductance (i.e., in inverse proportion to their resistance). Kirchhoff's current law (KCL) is the operative rule for parallel circuits. It states that the sum of all currents entering and exiting a node must sum to zero. Alternately, it can be stated as the sum of currents entering a node must equal the sum of currents exiting that node. As a pseudo formula:

$$\Sigma I \rightarrow = \Sigma I \leftarrow$$

It is possible to drive a parallel circuit with multiple current sources. These sources will add in much the same way that voltage sources in series add, that is, polarity must be considered. Ordinarily, voltage sources with differing values are not place in parallel as this violates the basic rule of parallel circuits (voltage being the same across all components).

When placing resistors in parallel it is good to remember that each additional resistor creates an extra path for current flow, thus increasing conductance and reducing total resistance. The total conductance equals the sum of the conductances of the individual resistors, $G_T = G_1 + G_2 + G_3 + ...$ or $1/R_T = 1/R_1 + 1/R_2 + 1/R_3 + ...$. Thus the total resistance is equal to the reciprocal of the sum of the reciprocals, $R_T = 1/(1/R_1 + 1/R_2 + 1/R_3 + ...)$. For two resistors, this formula can be rewritten as the "product-sum rule", or $R_T = (R_1 \cdot R_2)/(R_1 + R_2)$. If we think of R_2 as being a multiple, N, of R_1, or $N R_1$, this can be written as $R_T = R_1 \cdot N/(N + 1)$. For example, if the ratio of the two resistors is 3:1, then the parallel equivalent will be 3/4^{ths} of the smaller resistor. Parallel combinations of resistors *always* wind up being smaller than the smallest resistor in that group. The larger the resistor, the less impact it has on the total resistance.

Just as series circuits exhibit the voltage divider rule (voltage dividing in proportion to resistance), parallel circuits exhibit the *current divider rule* which states that current divides in reverse proportion to resistance (i.e., in direct proportion to conductance). This can be reduced to a simple formula when only two resistors are involved. Given two resistors, R_1 and R_2, and a current feeding them, I_T, the current through one of the resistors will equal the total current times the ratio of the *opposite* resistor over the sum of the two resistors. For example, $I_1 = I_T \cdot R_2/(R_1 + R_2)$. This rule is convenient in that you don't have to compute the parallel equivalent resistance, but remember, *it is valid only when there are just two resistors involved.* A more general version that can be used for any number of resistors is $I_i = I_T \cdot R_P/R_i$, where R_P is the total equivalent parallel resistance and R_i is the resistor of interest. This is, in essence, merely a rewriting of the fact that all components must see the same voltage: $I_i \cdot R_i = I_T \cdot R_P$.

When analyzing a parallel circuit, if it is being driven by a voltage source, then this same voltage must appear across each of the individual resistors. Ohm's law can then be used to determine the individual currents. According to KCL, the total current exiting the source must be equal to the sum of these individual currents. For example, in the circuit shown in Figure 3A, the voltage E must appear across both R1 and R2. Therefore, the currents must be I1 = E/R1 and I2 = E/R2, and Itotal = I1 + I2. Itotal can also be found by determining the parallel equivalent of R1 and R2 (usually written as R1 || R2), and then dividing this into E. This technique can also be used in reverse in order to determine a resistance value that will produce a given total current: dividing the source by the current yields the equivalent parallel resistance. As one of the two is already known, the parallel resistor can be used to determine the unknown resistor value).

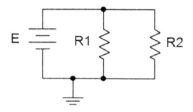

Figure 3A

If the parallel circuit is driven by a current source, as shown in Figure 3B, there are two basic methods of solving for the resistor currents. The fastest method is to simply use the current divider rule. If desired, the component voltage can then be found using Ohm's law. An alternate method involves finding the parallel equivalent resistance first, and then using Ohm's law to determine the voltage. Given the voltage, Ohm's law can be used to find the current through R1. To find the current through R2, either Ohm's law can be applied a second time, or KCL may be used, subtracting the current through R1 from the source current.

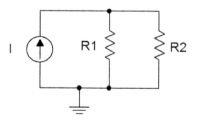

Figure 3B

It is worth noting that both methods described above will yield the correct answers. One is not "more correct" than the other. We can consider each of these as a separate "solution path"; that is, a method of arriving at the desired end point. In general, the more complex the circuit, the more solution paths there will be. This is good because one path may be more obvious to you than another. It also allows you a means of cross-checking your work.

3.1 Exercises

Analysis

1. Determine the effective resistance of the network shown in Figure 3.1.

Figure 3.1

2. Determine the effective resistance of the network shown in Figure 3.2.

Figure 3.2

3. Determine the effective resistance of the network shown in Figure 3.3.

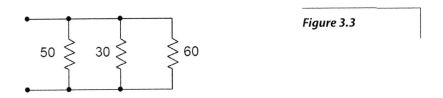

Figure 3.3

4. Find the effective source current of the network shown in Figure 3.4.

Figure 3.4

5. Determine the effective source current of the network shown in Figure 3.5.

Figure 3.5

6. Find the source and resistor currents for the circuit of Figure 3.6.

Figure 3.6

7. Determine the source and resistor currents for the circuit of Figure 3.7.

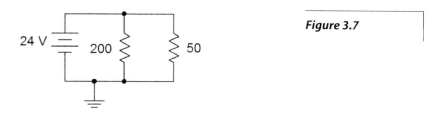

Figure 3.7

8. Find the source and resistor currents for the circuit of Figure 3.8.

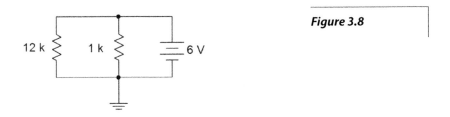

Figure 3.8

9. For the circuit of Figure 3.9, determine the source and resistor currents.

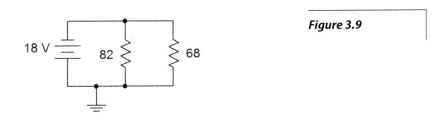

Figure 3.9

10. Determine the current through each resistor in the circuit of Figure 3.10. Also determine the total power generated by the source.

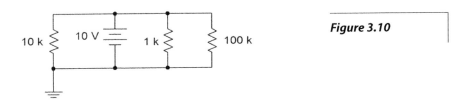

Figure 3.10

11. Consider the circuit shown in Figure 3.10. Assume that the 100 kΩ is replaced with another resistor ten times as large. Will this have a major impact on the current exiting source? Why/why not?

12. Consider the circuit shown in Figure 3.10. Assume that the 1 kΩ is replaced with another resistor ten times smaller. Will this have a major impact on the current exiting source? Why/why not?

13. Find the current through each resistor in the circuit of Figure 3.11.

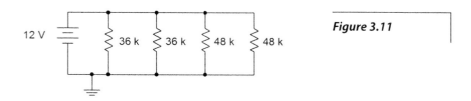

Figure 3.11

14. Determine the current through each resistor in the circuit of Figure 3.12. Also determine the total current exiting by the source.

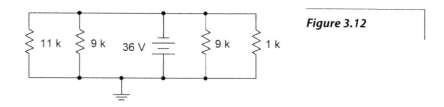

Figure 3.12

15. Determine the current through each resistor in the circuit of Figure 3.13.

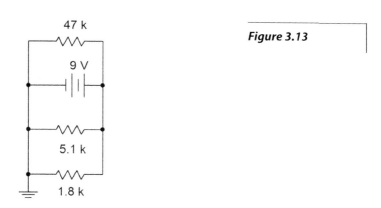

Figure 3.13

16. For the circuit shown in Figure 3.14, determine the current through each resistor and the source voltage.

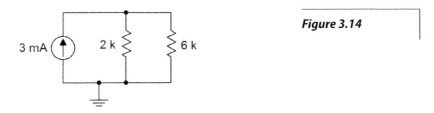

Figure 3.14

17. For the circuit shown in Figure 3.15, determine the current through each resistor and the source voltage.

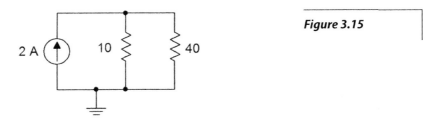

Figure 3.15

18. For the circuit shown in Figure 3.16, determine the current through each resistor and the source voltage.

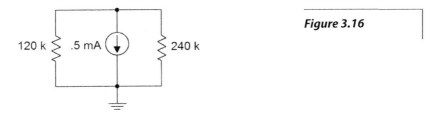

Figure 3.16

19. For the circuit shown in Figure 3.17, determine the current through each resistor and the source voltage.

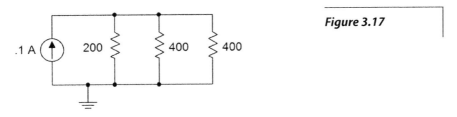

Figure 3.17

20. For the circuit shown in Figure 3.18, determine the current through each resistor and the source voltage.

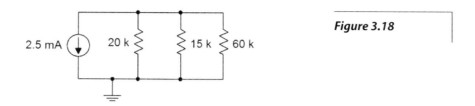

Figure 3.18

21. For the circuit shown in Figure 3.19, determine the current through each resistor and the source voltage.

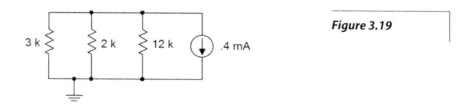

Figure 3.19

22. For the circuit shown in Figure 3.20, determine the current through each resistor and the source voltage.

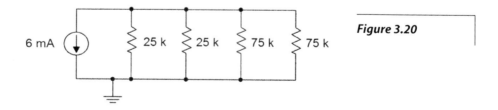

Figure 3.20

23. Referring to the circuit of Figure 3.20, determine the resistor currents if the right-most 75 kΩ resistor is accidentally opened (i.e., unconnected). How do these results compare to those of problem 22?

24. Referring to the circuit of Figure 3.20, determine the resistor currents if the right-most 75 kΩ resistor is accidentally shorted. How do these results compare to those of problems 22 and 23?

25. For the circuit shown in Figure 3.21, determine the current through each resistor and the source voltage.

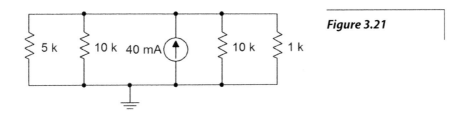

Figure 3.21

26. Given the circuit of Figure 3.22, find the currents through the two resistors.

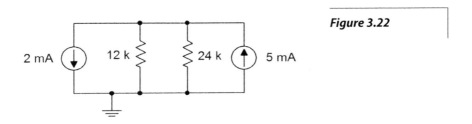

Figure 3.22

27. If the 5 mA current source shown in Figure 3.22 is accidentally wired in upside down, does the voltage across the 12 k Ω resistor become more positive or more negative with respect to ground?

28. Given the circuit of Figure 3.23, find the voltages across the three resistors.

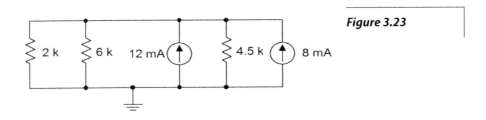

Figure 3.23

29. Find the currents through the three resistors in Figure 3.23.

41

Design

30. For the network shown in Figure 3.24, determine a for values for R1 given that R2 is 12 kΩ and the equivalent combination is 8 kΩ.

Figure 3.24

31. Add a third parallel resistor to the circuit of Figure 3.8 such that the source current is 10 mA.

32. Add a fourth parallel resistor to the circuit of Figure 3.10 such that the source current is 20 mA.

33. Consider the circuit shown in Figure 3.14. Determine a new value for the current source such that the source voltage equals 10 volts.

34. Consider the circuit shown in Figure 3.16. Determine a new value for the current source such that the source voltage equals 20 volts.

35. Given the circuit of Figure 3.25, if the source is 6 volts and R1 is 2 kΩ, what must be the value of R2 if the total current exiting the source is 10 mA?

Figure 3.25

36. For the circuit shown in Figure 3.26, determine values for resistors R2 and R3 such that the current through R2 is twice the current through R1 and the current through R3 is half the current through R1. The source is 6 volts and R1 is 2 kΩ.

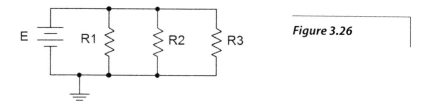

Figure 3.26

37. For the circuit shown in Figure 3.27, determine values for resistors R1 and R2 such that the current through R2 is twice the current through R1. The source is 10 mA and R1 is 6 kΩ.

Figure 3.27

Challenge

38. For the circuit shown in Figure 3.12, determine a new value for the 11 kΩ resistor such that the supply current is 50 mA.

39. For the circuit shown in Figure 3.14, determine a new value for the 2 kΩ resistor such that the voltage drop across the 6 kΩ is 15 volts.

40. Consider the circuit shown in Figure 3.21. If the current source was replaced with a voltage source, what value is needed so that the same currents flow through the resistors as in the original circuit?

41. For the network shown in Figure 3.24, determine values for R1 and R2 such that R2 is twice the size of R1 and the equivalent combination is 6 kΩ.

42. Given the network of Figure 3.3, is it possible to replace the 60 Ω resistor with another value such that the equivalent combination of the three resistors is 25 Ω? If so, what is that value?

43. Given the network of Figure 3.11, is it possible to add a fifth parallel resistor such that the source current is 1 mA? If so, what is that value?

44. For the circuit shown in Figure 3.26, determine values for the three resistors such that the current through R1 is twice the current through R2 and four times the current through R3. The source is 12 volts and should produce a total of 9 mA of current.

45. For the circuit shown in Figure 3.27 determine values for the two resistors such that the current through R1 is half the current through R2. The source is 24 mA and should produce a drop of 16 volts across R1.

46. Given three current sources with values of 1 mA, 2 mA and 7 mA; how would they need to be connected in order to deliver 4 volts across a 1 kΩ load resistor?

47. Consider the circuit of Figure 3.28. Assume I is a 4 mA source. Using only 5% standard resistor values (see Appendix A), pick values for the three resistors such that the voltage across R1 is within 10% of 10 volts as long as the resistors are within tolerance.

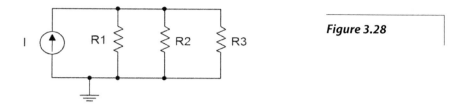

Figure 3.28

Simulation

48. Verify the currents found in problem 11 via a DC simulation.

49. Verify the currents found in problem 15 via a DC simulation.

50. Verify the currents and voltages found in problem 17 via a DC simulation.

51. Verify the results found in problem 25 via a DC simulation.

52. Verify the results found in problem 27 via a DC simulation.

53. Perform a DC simulation on the design of problem 44 to verify its performance.

54. Perform a DC simulation on the design of problem 45 to verify its performance.

55. Perform a DC simulation on the design of problem 46 to verify its performance.

56. Perform a Monte Carlo or worst-case simulation on the design of problem 47 to verify its performance.

4 Series-Parallel Resistive Circuits

This section covers:
- Circuits using multiple resistors in series-parallel with either a single voltage source or a current source.

4.0 Introduction

There is an infinite variety of series-parallel circuits. This section deals with a subset, namely those that are driven by a single current source or voltage, and which may be simplified using series and parallel resistor combinations. The key to analyzing series-parallel circuits is in recognizing portions of the circuit that are in series or in parallel and then applying the series and parallel analysis rules to those sections. Ohm's law, KVL and KCL may be used in turn to "chip away" at the problem until all currents and voltages are found. As individual voltages and currents are determined, this makes it easier to apply these rules to determine other values.

Consider the circuit of Figure 4A. This is neither just a series circuit nor just a parallel circuit. If it was a series circuit then the current through all components would have to be same, that is, there would no nodes where the current could divide. This is clearly not the case as the current flowing through R1 can divide at node b, with one portion flowing down through R2 and the remainder through R3. On the other hand, if it was strictly parallel, then all of the components would have to exhibit the same voltage and therefore there would be only two connection points in the circuit. This is also not the case as there are three such points: a, b and ground.

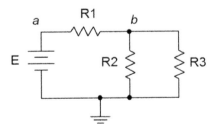

Figure 4A

What is true is that resistors R2 and R3 are in parallel. We know this because both components are attached to the same two nodes, b and ground, and must exhibit the same voltage, V_b. As such, we can find the equivalent resistance of this pair and treat the result as a single resistance, let's call it Rx. In this newly simplified circuit, Rx is in series with R1 and the source, E. We have simplified the original circuit into a simple series circuit and thus the series analysis rules may be applied.

There are many solution paths at this point. For example, we could find the total resistance, Rt, by adding R1 to Rx. Dividing this by E yields the total current flowing out of the source, I_{total}. This current must flow through R1 so Ohm's law can be used to find the voltage drop across R1. This same current must be flowing through Rx, so Ohm's law can be used to find the associated voltage (V_b). The currents through R2 and R3 may then be found using Ohm's law for each resistor (e.g., the current through R2 must be V_b/R2. Alternately, these currents may be found by using the current divider rule between R2 and R3 (e.g., the current through R2 must be I_{total} · R1/(R1 + R2); remember, the current divider rule uses the ratio of the *opposite* resistor over the sum).

Another solution path would be to apply the voltage divider rule to R1 and Rx in order to derive the two voltage drops (or the rule can be applied to find just one of the drops and the other voltage may be found by subtracting that from the source, an application of KVL). Once the voltages are determined, Ohm's law can be used to find the currents.

As the circuit grows, more and more solution paths exist. Consider the circuit of Figure 4B. In this case, R3 and R4 are in parallel. This parallel combination is in series with R2. Finally, this set of three resistors is in parallel with R1 and E, reducing to a parallel circuit. Consequently, we know that the voltage across R1 must be equal to E. Also, the currents through R1 and R2 must add up to the current exiting the source (KCL). Further, the currents through R3 and R4 must add up to the current flowing through R2 (KCL). Also, the voltages across R3 and R4 must be the same (they are in parallel) and that this voltage plus the voltage across R2 must equal E (KVL). One solution path would be to find the total resistance as seen by the source, R1 ∥ (R2 + (R3 ∥ R4)), and use this to find the total current flowing out of the source. The current divider rule can then be used between R1 and (R2 + (R3 ∥ R4)). Ohm's law can be used subsequently to find various voltage drops. Alternately, the voltage divider rule can be used between R2 and (R3 ∥ R4) as this combination is driven by E. Knowing the voltages, the currents may be determined. With so many possible solution paths for large circuits, it is often worthwhile to take a moment to map out a strategy rather than just "diving in" and hoping it will all work out.

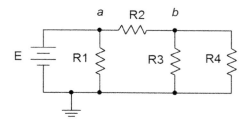

Figure 4B

Series-parallel simplification techniques will not work for all circuits. Some resistive networks such as delta or bridge configurations require other techniques that will be addressed in later sections.

4.1 Exercises

Analysis

1. In the circuit of Figure 4.1, which individual resistors are strictly in series and which are in parallel?

Figure 4.1

2. In the circuit of Figure 4.2, which individual resistors are strictly in series and which are in parallel?

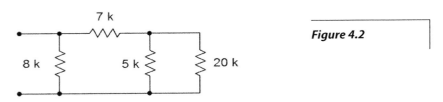

Figure 4.2

3. In the circuit of Figure 4.3, which individual resistors are strictly in series and which are in parallel?

Figure 4.3

4. In the circuit of Figure 4.4, which individual resistors are strictly in series and which are in parallel?

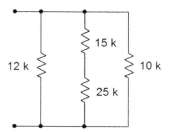

Figure 4.4

5. In the circuit of Figure 4.5, which individual resistors are strictly in series and which are in parallel?

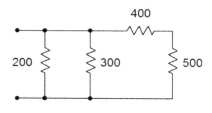

Figure 4.5

6. In the circuit of Figure 4.6, which individual resistors are strictly in series and which are in parallel?

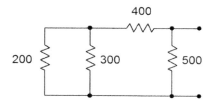

Figure 4.6

7. In the circuit of Figure 4.7, which individual resistors are strictly in series and which are in parallel?

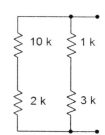

Figure 4.7

8. In the circuit of Figure 4.8, which individual resistors are strictly in series and which are in parallel?

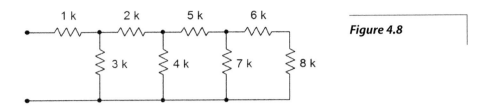

Figure 4.8

9. Determine the equivalent resistance of the network shown in Figure 4.1 (i.e., as if an ohmmeter is connected to the two terminals).

10. Determine the equivalent resistance of the network shown in Figure 4.2.

11. Determine the equivalent resistance of the network shown in Figure 4.3.

12. Determine the equivalent resistance of the network shown in Figure 4.4.

13. Determine the equivalent resistance of the network shown in Figure 4.5.

14. Determine the equivalent resistance of the network shown in Figure 4.6.

15. Determine the equivalent resistance of the network shown in Figure 4.7.

16. Determine the equivalent resistance of the network shown in Figure 4.8.

17. For the circuit of Figure 4.9, find voltages V_a, V_b and V_{ab}.

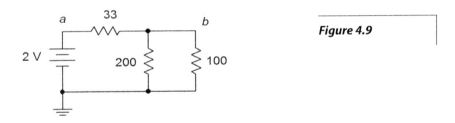

Figure 4.9

18. For the circuit of Figure 4.9, find the current through each resistor.

19. For the circuit of Figure 4.10, find the current through each resistor.

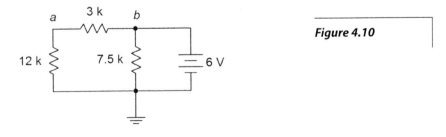

Figure 4.10

20. For the circuit of Figure 4.10, find voltages V_a, V_b and V_{ab}.

21. For the circuit of Figure 4.11, find voltages V_a, V_b and V_{ab}.

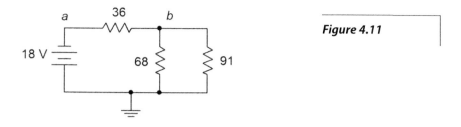

Figure 4.11

22. For the circuit of Figure 4.11, find the current through each resistor.

23. For the circuit of Figure 4.12, find the current through each resistor.

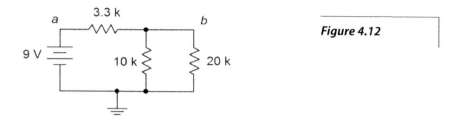

Figure 4.12

24. For the circuit of Figure 4.12, find voltages V_a, V_b and V_{ab}.

25. In the circuit of Figure 4.13, find voltages V_a, V_b and V_{ab}.

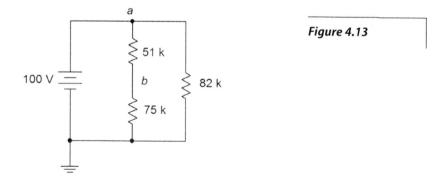

Figure 4.13

26. In the circuit of Figure 4.13, find the current through each resistor.

27. In the circuit of Figure 4.14, find the current through each resistor.

28. In the circuit of Figure 4.14, find voltages V_a, V_b and V_{ab}.

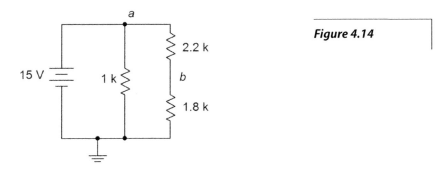

Figure 4.14

29. For the circuit of Figure 4.15, find voltages V_b, V_c and V_{cb}.

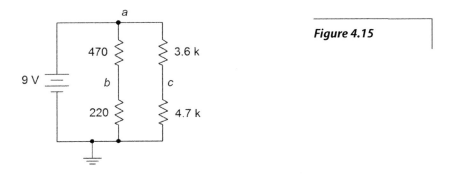

Figure 4.15

30. For the circuit of Figure 4.15, find the current through the 470 Ω and 3.6 kΩ resistors along with the source current.

31. For the circuit of Figure 4.16, find the current through the 2 kΩ and 5.1 kΩ resistors along with the source current.

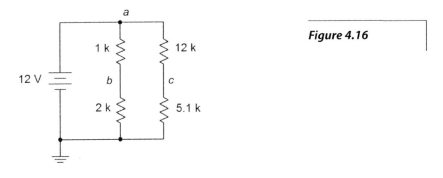

Figure 4.16

32. For the circuit of Figure 4.16, find voltages V_c, V_b, V_{bc} and V_{ab}.

33. For the circuit of Figure 4.17, find voltages V_a and V_b.

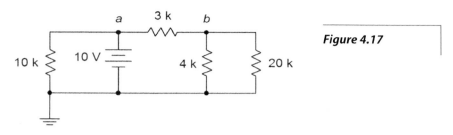

Figure 4.17

34. For the circuit of Figure 4.17, find the current through the 10 kΩ and 4 kΩ resistors along with the source current.

35. For the circuit of Figure 4.18, find the current through the 30 kΩ and the left-most 36 kΩ resistors along with the source current.

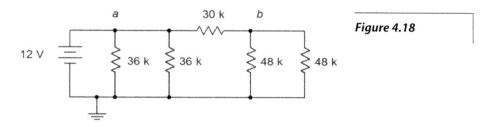

Figure 4.18

36. For the circuit of Figure 4.18, find voltages V_a, V_b and V_{ab}.

37. In the circuit of Figure 4.18, must the currents through the two 36 kΩ resistors be the same? Why/why not?

38. In the circuit of Figure 4.18, must the currents through the two 48 kΩ resistors be the same? Further, must these currents be the same as the currents through the two 36 kΩ resistors Why/why not?

39. For the circuit of Figure 4.19, find voltages V_a, V_b and V_{ab}.

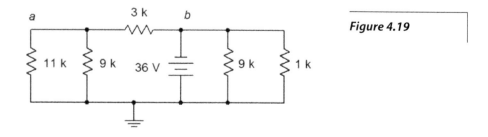

Figure 4.19

40. For the circuit of Figure 4.19, find the current through the 3 kΩ, 11 kΩ, and both 9 kΩ resistors.

41. Given the circuit of Figure 4.20, find the current through the 47 kΩ, 5.1 kΩ, and 3.9 kΩ resistors.

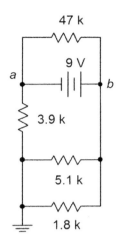

Figure 4.20

42. Given the circuit of Figure 4.20, find voltages V_a, V_b and V_{ab}.

43. For the circuit of Figure 4.21, find voltages V_b, V_c and V_d.

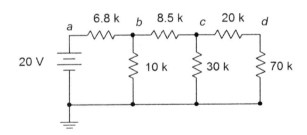

Figure 4.21

44. For the circuit of Figure 4.21, find the current through the 6.8 kΩ, 8.5 kΩ, and 20 kΩ resistors.

45. The circuit of Figure 4.22 is called an R-2R ladder. Find the current through each of the 60 kΩ resistors.

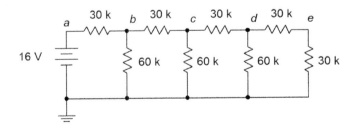

Figure 4.22

46. Given the circuit of Figure 4.22, find voltages V_b, V_c, V_d and V_e. What is unique about this configuration?

47. For the circuit of Figure 4.23, find voltages V_a, V_b and V_{ab}.

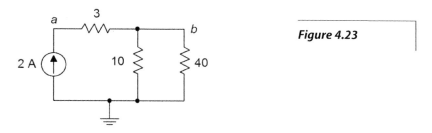

Figure 4.23

48. For the circuit of Figure 4.23, find the current through each of the resistors.

49. For the circuit of Figure 4.24, find the current through each of the resistors.

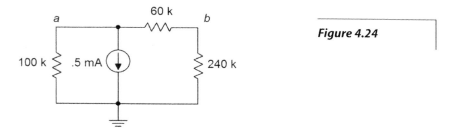

Figure 4.24

50. For the circuit of Figure 4.24, find voltages V_a, V_b and V_{ab}.

51. Given the circuit of Figure 4.25, find voltages V_a, V_b and V_{ab}.

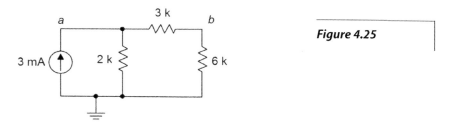

Figure 4.25

52. Given the circuit of Figure 4.25, find the current through each of the resistors.

53. For the circuit of Figure 4.26, find the current through the 200 Ω, 400 Ω, and 100 Ω resistors.

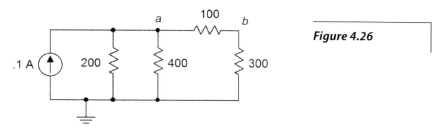

Figure 4.26

54. For the circuit of Figure 4.26, find voltages V_a, V_b and V_{ab}.

55. Given the circuit of Figure 4.27, find voltages V_a, V_b and V_{ab}.

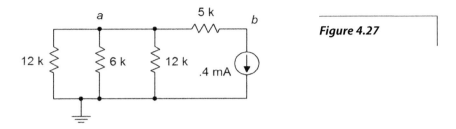

Figure 4.27

56. Given the circuit of Figure 4.27, find the current through the 5 kΩ, and 6 kΩ resistors.

57. For the circuit of Figure 4.28, find the current through the 9 kΩ and right-most 82 kΩ resistors.

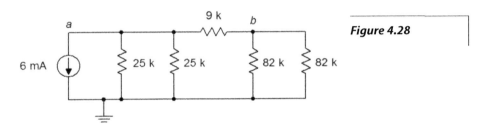

Figure 4.28

58. For the circuit of Figure 4.28, find voltages V_a, V_b and V_{ab}.

59. Must the current through the 82 kΩ resistors be identical in the circuit of 4.28? Why/why not?

60. Must the current through the 25 kΩ resistors be identical in the circuit of 4.28? Must they be the same as the currents through the 82 kΩ resistors? Why/why not?

61. For the circuit of Figure 4.29, find voltages V_a, V_b and V_{ab}.

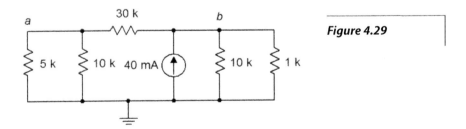

Figure 4.29

62. For the circuit of Figure 4.29, find the current through the 30 kΩ, 5 kΩ, and 1 kΩ resistors.

63. For the circuit of Figure 4.30, find the current through the 1 kΩ, 2.2 kΩ, and 18 kΩ resistors.

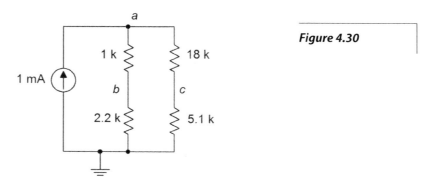

Figure 4.30

64. For the circuit of Figure 4.30, find voltages V_a, V_b and V_c.

65. Given the circuit of Figure 4.31, find voltages V_a, V_b and V_{ab}.

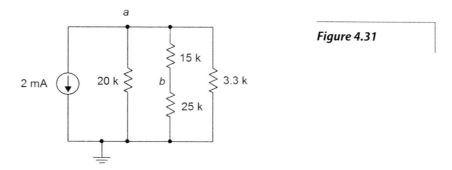

Figure 4.31

66. Given the circuit of Figure 4.31, find the current through the 20 kΩ, 15 kΩ, and 3.3 kΩ resistors.

67. The circuit of Figure 4.32 is referred to as an R-2R ladder network. Find the current through each of the 4 kΩ resistors.

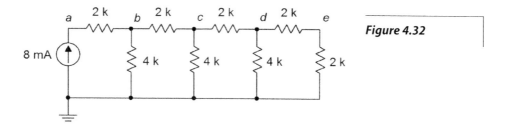

Figure 4.32

68. For the circuit of Figure 4.32, find voltages V_a, V_b and V_{ab}. What is the unique characteristic of this configuration?

Design

69. Determine a new value for the 33 Ω resistor in Figure 4.9 such that the source current is 10 mA.
70. Determine a new value for the 20 kΩ resistor in Figure 4.12 such that V_b is 4 volts.
71. Determine a new value for the 470 Ω resistor in Figure 4.15 such that V_b is 6 volts.
72. Determine a new value for the 1 kΩ resistor in Figure 4.14 such that the source current is 20 mA.
73. Determine a new value for the 12 kΩ resistor in Figure 4.16 such that V_{bc} is 0 volts.
74. Given the circuit of Figure 4.24, determine a new value for the 100 kΩ resistor such that V_a is 75 volts.
75. Given the circuit of Figure 4.23, determine a new value for the current source such that V_b is 8 volts.
76. Determine a new value for the 2.2 kΩ resistor in Figure 4.30 such that V_{bc} is 0 volts.
77. Given the circuit of Figure 4.32, determine a new value for the current source such that V_e is 1 volt.

Challenge

78. Utilizing only 1 kΩ resistors, create a series-parallel combination that achieves 1.25 kΩ of total resistance.
79. Utilizing only 12 kΩ resistors, create a series-parallel combination that achieves 9 kΩ of total resistance.
80. Consider the circuit of Figure 4.22. Alter the values of the two right-most 30 kΩ resistors such that V_e is 1.2 volts. The remaining node voltages should be unchanged from the original circuit.
81. Alter the circuit shown in Figure 4.22 such that another "wrung" is added to the ladder creating a new right-most node f. If the former terminating 30 kΩ resistor (i.e., the vertical one) is reset to 60 kΩ, determine the values of the resistors on the new wrung such that V_f is 0.5 volts.
82. Given the circuit of Figure 4.25, determine a new value for the 2 kΩ resistor such that V_b is 12 volts.
83. Given the circuit of Figure 4.32, determine new values for the resistors such that all of the node voltages are twice the value of the original circuit's node voltages.

Simulation

84. Perform a DC simulation on the result of problem 24 to verify the node voltages.
85. Perform a DC simulation on the result of problem 25 to verify the node voltages.
86. Perform a DC simulation on the result of problem 43 to verify the node voltages.
87. Perform a DC simulation on the result of problem 46 to verify the node voltages.
88. Perform a DC simulation on the result of problem 51 to verify the node voltages.
89. Perform a DC simulation on the result of problem 68 to verify the node voltages.
90. Perform a DC simulation on the proposed solution to problem 69 to verify the new design.
91. Perform a DC simulation of the alteration presented in Challenge problem 80. Does the new design meet all of the requirements?
92. Perform a DC simulation of the alteration requested in Challenge problem 83. Does the new design meet all of the requirements?

5 Analysis Theorems and Techniques

This section covers:
- Superposition theorem for multi-source circuits.
- Source conversions.
- Thévenin's theorem.
- Norton's theorem.
- Maximum power transfer theorem.
- Delta-Y conversions.

5.0 Introduction

Superposition Theorem

Superposition allows the analysis of multi-source series-parallel circuits. Superposition can only be applied to networks that are linear and bilateral. Further, it cannot be used to find values for non-linear functions, such as power, directly (although power can be computed from the resulting voltage or current values). The basic idea is to determine the contribution of each source by itself, and then adding the results to get the final answer(s). Consider the circuit depicted in Figure 5A, below.

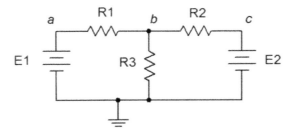

Figure 5A

Here we see two voltage sources, E1 and E2, driving a three resistor series-parallel network. As there are two sources, two derived circuits must be created; one using only E1 and the other using only E2. When considering a given source, all other sources are replaced by their ideal internal resistance. In the case of a voltage source, that's a short; and in the case of a current source, that's an open. When considering E1, E2 is replaced with a short. This leaves is a fairly simple network where R3 and R2 are in parallel. This combination is in series with R1 and E1. Using basic series-parallel techniques, we can solve for desired quantities such as the current flowing through R1 or the voltage V_b. Be sure to indicate the current direction and voltage polarity (here, that's left-to-right and positive). The process is then repeated for E2, shorting E1 and leaving us with R1 in parallel with R3 which is in turn in series with R2 and E2. Note that although in this version V_b is still positive, the current direction for R1 is now right-to-left. The numerical results from this version are added to those of the E1 version (minding polarities and directions) to achieve the final result. If power is needed, it can be computed from these currents and voltages. Note that superposition can work with a mix of current sources and voltage sources. The practical downside is that for large circuits using many sources, many derived circuits will need to be analyzed. For example, if there are three voltage sources and two current sources, then a total of five derived circuits will be created.

Source Conversions

For any simple voltage source consisting of an ideal voltage source with a series internal resistance, an equivalent current source may be created. Similarly, for any simple current source consisting of an ideal current source with a parallel internal resistance, an equivalent voltage source may be created. By "equivalent", we mean that both circuits will produce the same voltage and current to identical loads. Consider the simple voltage source of Figure 5B. It's equivalent current source is shown in Figure 5C.

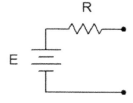

Figure 5B

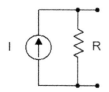

Figure 5C

For reasons that will become apparent under the section on Thévenin's theorem below, the internal resistances of these two circuits must be identical if they are to behave identically. Knowing that, it is a straight-forward process to find the required value of the other source. For example, given a voltage source, the maximum current that can be developed occurs when the load is shorted. This current is E/R. Under that same load condition, all of the current from the current source version must be flowing through the load. Therefore, the value of the equivalent current source must be the maxim current of E/R. Similarly, if we start with a current source, an open load produces the maximum load voltage of I · R.

Therefore, the equivalent voltage source must have a value of I · R. Finally, if a multi-source is being converted (i.e., voltage sources in series or current sources in parallel), first combine the sources to arrive at the simplest source and then do the conversion. Do not convert the sources first and then combine them as you will wind up with series-parallel configurations rather than simple sources.

Judicious use of source conversions can sometimes simplify multi-source circuits by allowing converted sources to be combined, resulting in a single source.

Thévenin's Theorem

Thévenin's theorem states that any two point linear network can be reduced to a simple voltage source (Eth) in series with an internal resistance (Rth) as shown in Figure 5D. This is a powerful analysis tool.

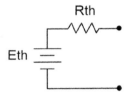

Figure 5D

The phrase "two point network" means that the circuit is cut in such a way that only two connections exist to the remainder of the circuit (i.e., one port). That remainder may be a single component or a large multi-component sub-circuit. As there are many ways to cut a typical circuit, there are many possible Thévenin equivalents. Consider the circuit shown in Figure 5E.

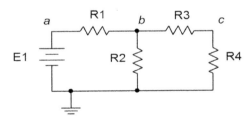

Figure 5E

Suppose we cut the circuit immediately to the left of R4. That is, we will find the Thévenin equivalent that drives R4. The first step is to make the cut, removing the remainder of the circuit (in this case, just R4). We then determine the open circuit output voltage. This is the maximum voltage that could appear between the cut points and is called the Thévenin voltage, Eth. This is shown in Figure 5F, following. In a circuit such as this, basic series-parallel analysis may be used to find Eth (note that due to the open, no current flows through R3, thus no voltage is developed across R3, and therefore Eth must equal the voltage developed across R2).

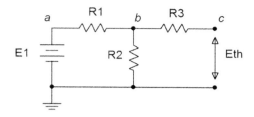

Figure 5F

The second part is finding the Thévenin resistance, Rth. Beginning with the "cut" circuit, replace all sources with their ideal internal resistance (thus shorting voltage sources and opening current sources). From the perspective of the cut point, look back into the circuit and simplify to determine its equivalent resistance. This is shown in Figure 5G. Looking in from where the cut was made (right-to-left), we find that R1 and R2 are in parallel, and this combination is then in series with R3. Thus, Rth = R3 + (R1 || R2).

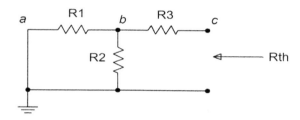

Figure 5G

As noted earlier, the original circuit could be cut in a number of different ways. We might, for example, want to determine the Thévenin equivalent that drives R2 in the circuit above. The cut appears below in Figure 5H.

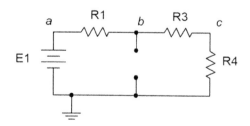

Figure 5H

Clearly, this will result in different values for both Eth and Rth. For example, Rth is now R1 || (R3 + R4).

Norton's Theorem

In a nutshell, Norton's theorem is the current source version of Thévenin's theorem. That is, a two point network can be reduced to a single current source with parallel internal resistance. The process is very similar. First, the Norton resistance is the same as the Thévenin resistance. Second, instead of finding the open circuit output voltage, you find the short circuit output current (again, the maximum value). This is the Norton current. If you can create a Thévenin equivalent for a network, then it must be possible to create a Norton equivalent. Indeed, if a Thévenin equivalent is found, a source conversion can be performed on it to yield the Norton equivalent.

Maximum Power Transfer Theorem

The maximum power transfer theorem states that in order to achieve the maximum power in a load, the load resistance must be equal to the internal resistance of the source. No other value of load resistance will produce a higher load power. For the circuit of Figure 5I, this means that RL must equal Ri.

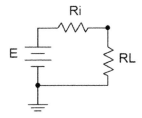

Figure 5I

While this produces the maximum load power, it does not produce maximum load current or maximum load voltage. In fact, this condition produces a load voltage and current that are half of their maximums. Their product, however, is at the maximum. Further, efficiency at maximum load power is only 50%. Values of RL greater than Ri will achieve higher efficiency but at reduced load power.

As any linear two point network can be reduced to something like Figure 5I by using Thévenin's theorem, combining the two theorems allows you to determine maximum power conditions for any resistor in a complex circuit.

Delta-Y Conversions

Certain component configurations, such as bridged networks, cannot be reduced to a single resistance using basic series-parallel conversion techniques. One method for simplification involves converting sections into more convenient forms. The configurations in question are three point networks containing three resistors. Due to the manner in which they drawn, they are referred to as delta (Δ) networks and Y networks. Alternately, if they are slightly redrawn they are known as pi (π) networks and T networks. These networks are shown in Figures 5J and 5K.

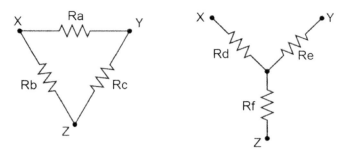

Figure 5J, Delta-Y (Δ-Y)

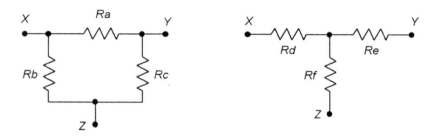

Figure 5K, Pi-T (π-T)

It is possible to convert back and forth between delta and Y networks. That is, for every delta network, there exists a Y network such that the resistances seen between the X, Y and Z terminals are identical, and vice versa. Consequently, one configuration can replace another in order to simplify a larger circuit.

To convert from delta to Y:

Rd = (Ra·Rb)/(Ra+Rb+Rc)
Re = (Ra·Rc)/(Ra+Rb+Rc)
Rf = (Rb·Rc)/(Ra+Rb+Rc)

To convert from Y to delta:

Ra = (Rd·Re+Re·Rf+Rd·Rf)/(Rf)
Rb = (Rd·Re+Re·Rf+Rd·Rf)/(Re)
Rc = (Rd·Re+Re·Rf+Rd·Rf)/(Rd)

5.1 Exercises

Analysis

1. For the circuit shown in Figure 5.1, determine the equivalent current source.

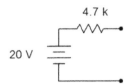
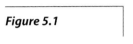

Figure 5.1

2. Given the circuit shown in Figure 5.2, determine the equivalent current source.

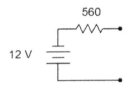
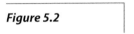

Figure 5.2

3. Determine the equivalent current source for the circuit shown in Figure 5.3.

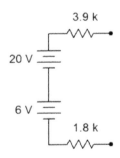
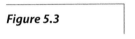

Figure 5.3

4. For the circuit shown in Figure 5.4, determine the equivalent current source.

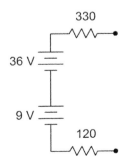
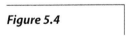

Figure 5.4

5. For the circuit shown in Figure 5.5, determine the equivalent voltage source.

Figure 5.5

6. Given the circuit shown in Figure 5.6, determine the equivalent voltage source.

Figure 5.6

7. Determine the equivalent voltage source for the circuit shown in Figure 5.7.

Figure 5.7

8. For the circuit shown in Figure 5.8, determine the equivalent voltage source.

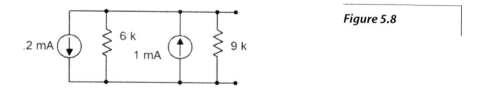

Figure 5.8

9. Given the circuit shown in Figure 5.9, determine the equivalent voltage source.

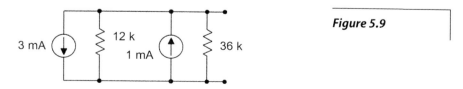

Figure 5.9

10. Using source conversion, find V_b for the circuit shown in Figure 5.10.

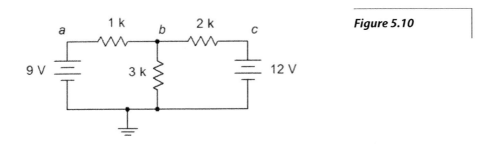

Figure 5.10

11. Using source conversion, find the current through the 3 kΩ resistor in the circuit of Figure 5.11.

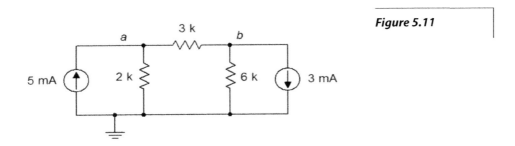

Figure 5.11

12. Using source conversion, find V_b for the circuit shown in Figure 5.12.

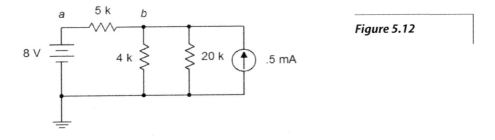

Figure 5.12

13. Using source conversion, find V_a for the circuit shown in Figure 5.13.

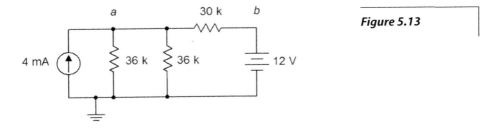

Figure 5.13

14. Using source conversion, find V_b for the circuit shown in Figure 5.14.

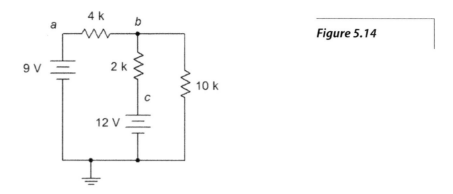

Figure 5.14

15. Using superposition, determine V_b for the circuit shown in Figure 5.10.
16. Using superposition, find the current through the 3 kΩ resistor for the circuit of Figure 5.10.
17. Using superposition, find the current through the 3 kΩ resistor for the circuit of Figure 5.11.
18. Using superposition, determine V_{ab} for the circuit shown in Figure 5.11.
19. Using superposition, determine V_b for the circuit shown in Figure 5.12.
20. Using superposition, find the current through the 4 kΩ resistor for the circuit of Figure 5.12.
21. Using superposition, find the current through the 30 kΩ resistor for the circuit of Figure 5.13.
22. Using superposition, determine V_a for the circuit shown in Figure 5.13.
23. Using superposition, determine V_{ba} for the circuit shown in Figure 5.14.
24. Using superposition, find the current through the 10 kΩ resistor for the circuit of Figure 5.14.
25. Using superposition, find the current through the 1.5 kΩ resistor for the circuit of Figure 5.15.

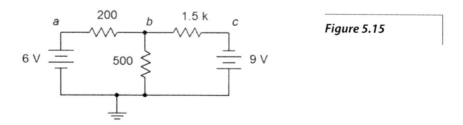

Figure 5.15

26. Using superposition, determine V_{ab} for the circuit shown in Figure 5.15.

27. Using superposition, determine V_b for the circuit shown in Figure 5.16.

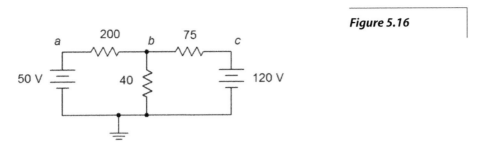

Figure 5.16

28. Using superposition, find the current through the 200 Ω resistor for the circuit of Figure 5.16.

29. Using superposition, find the current through the 4 kΩ resistor for the circuit of Figure 5.17.

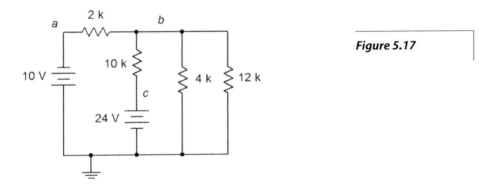

Figure 5.17

30. Using superposition, determine V_b for the circuit shown in Figure 5.17.

31. Is it possible to determine V_b in Figure 5.17 by using source conversions instead of superposition? Why/why not?

32. Using superposition, determine V_{bc} for the circuit shown in Figure 5.18.

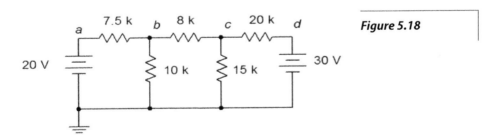

Figure 5.18

33. Using superposition, find the current through the 10 kΩ resistor for the circuit of Figure 5.18.

34. Using superposition, find the currents through the 100 Ω and 700 Ω resistors for the circuit shown in Figure 5.19.

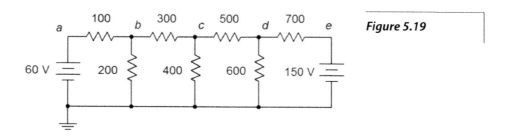

Figure 5.19

35. Using superposition, determine V_{bd} for the circuit shown in Figure 5.19.

36. Is it possible to determine V_{bd} in Figure 5.19 by using source conversions instead of superposition? Why/why not?

37. Using superposition, determine V_{ad} for the circuit shown in Figure 5.20.

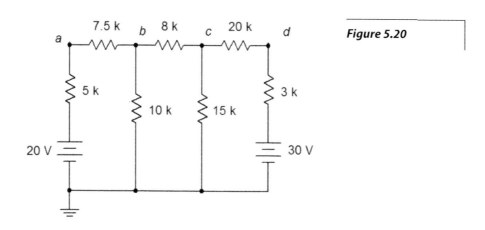

Figure 5.20

38. Using superposition, find the current through the 20 kΩ resistor for the circuit shown in Figure 5.20.

39. Using superposition, find the current through the 12 kΩ resistor for the circuit shown in Figure 5.21.

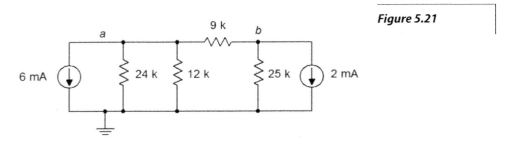

Figure 5.21

40. Using superposition, determine V_b for the circuit shown in Figure 5.21.

41. Using superposition, determine V_b for the circuit shown in Figure 5.22.

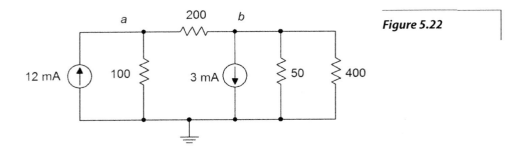

Figure 5.22

42. Using superposition, find the current through the 100 Ω resistor for the circuit of Figure 5.22.

43. Using superposition, find the current through the 5 kΩ resistor for the circuit of Figure 5.23.

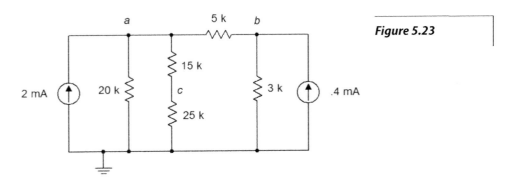

Figure 5.23

44. Using superposition, determine V_c for the circuit shown in Figure 5.23.

45. Given the circuit shown in Figure 5.24, determine the Thévenin equivalent circuit that is driving the 4 kΩ resistor.

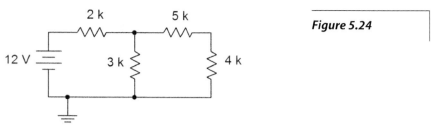

Figure 5.24

46. Given the circuit shown in Figure 5.24, determine the Norton equivalent circuit driving the 4 kΩ resistor.

47. Given the circuit shown in Figure 5.24, determine the Norton equivalent circuit driving the 12 kΩ resistor.

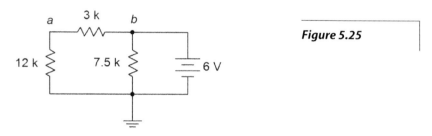

Figure 5.25

48. Determine the Thévenin equivalent circuit driving the 12 kΩ resistor for the circuit shown in Figure 5.25.

49. Given the circuit shown in Figure 5.26, determine the Thévenin equivalent circuit that is driving the 20 kΩ resistor.

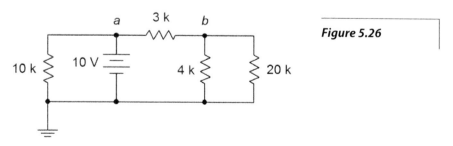

Figure 5.26

50. For the circuit shown in Figure 5.26, determine the Norton equivalent circuit driving the 4 kΩ resistor.

51. Given the circuit shown in Figure 5.27, determine the Norton equivalent circuit driving the 40 Ω resistor.

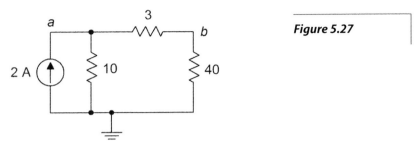

Figure 5.27

52. Determine the Thévenin equivalent circuit driving the 10 Ω resistor for the circuit shown in Figure 5.27.

53. Given the circuit shown in Figure 5.28, determine the Norton equivalent circuit driving the 12 kΩ resistor.

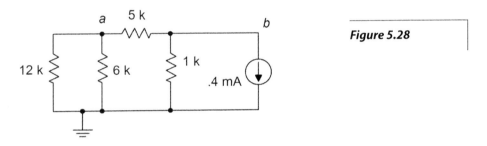

Figure 5.28

54. Given the circuit of Figure 5.24, determine the power in the 4 kΩ resistor. If this resistor can be replaced with any other value, is it possible to achieve a higher power? Why/why not?

55. Given the circuit of Figure 5.25, determine the power in the 12 kΩ resistor. If this resistor can be replaced with any other value, is it possible to achieve a higher power? Why/why not?

56. Given the circuit of Figure 5.27, determine the power in the 40 Ω resistor. If this resistor can be replaced with any other value, is it possible to achieve a higher power? Why/why not?

57. Given the circuit of Figure 5.28, determine the power in the 6 kΩ resistor. If this resistor can be replaced with any other value, is it possible to achieve a higher power? Why/why not?

Design

58. Consider the 4 kΩ resistor to be the load in Figure 5.24. Determine a new value for the load in order to achieve maximum load power. Also determine the maximum load power.

59. Consider the 12 kΩ resistor to be the load in Figure 5.25. Determine a new value for the load in order to achieve maximum load power. Also determine the maximum load power.

60. Consider the 40 Ω resistor to be the load in Figure 5.27. Determine a new value for the load in order to achieve maximum load power. Also determine the maximum load power.

61. Consider the 6 kΩ resistor to be the load in Figure 5.28. Determine a new value for the load in order to achieve maximum load power. Also determine the maximum load power.

62. Redesign the circuit of Figure 5.15 so that it uses only current sources and produces the same component currents and voltages as the original circuit.

63. Redesign the circuit of Figure 5.17 so that it uses only current sources and produces the same component currents and voltages as the original circuit.

64. Redesign the circuit of Figure 5.21 so that it uses only voltage sources and produces the same component currents and voltages as the original circuit.

65. Redesign the circuit of Figure 5.23 so that it uses only voltage sources and produces the same component currents and voltages as the original circuit.

66. Convert the delta network of Figure 5.29 into a Y network.

Figure 5.29

67. Convert the pi network of Figure 5.30 into a T network.

Figure 5.30

68. Convert the Y network of Figure 5.31 into a delta network.

Figure 5.31

69. Convert the T network of Figure 5.32 into a pi network.

Figure 5.32

Challenge

70. Redesign the circuit of Figure 5.33 so that it uses only current sources and produces the same node voltages as the original circuit.

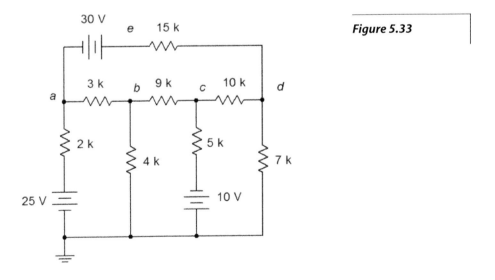

Figure 5.33

71. Using any combination of techniques, find the current through the 9 kΩ resistor for the circuit of Figure 5.33.

72. Using any combination of techniques, determine V_{bc} for the circuit shown in Figure 5.33.

73. Is it possible to determine V_{bc} in Figure 5.33 by using just source conversions? Why/why not?

74. Using superposition, find the current through the 25 Ω resistor for the circuit of Figure 5.34.

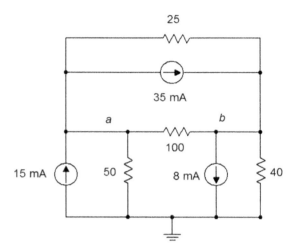

Figure 5.34

75. Using superposition, find V_{ab} in the circuit of Figure 5.34.

76. Redesign the circuit of Figure 5.34 using only voltage sources so that it achieves the same node voltages as the original.

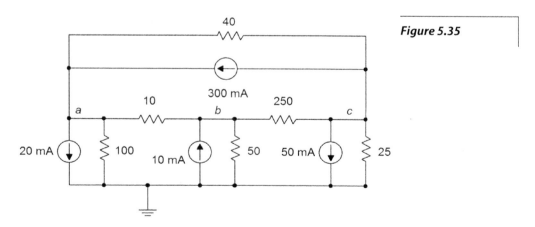

Figure 5.35

77. Is it possible to determine V_{bc} in Figure 5.35 by using just source conversions or just superposition? Why/why not?

78. Using any combination of techniques, find the current through the 100 Ω resistor for the circuit shown in Figure 5.35.

79. Redesign the circuit of Figure 5.35 so that it uses only voltage sources and produces the same node voltages as the original circuit.

80. Determine the Thévenin and Norton equivalents driving the 40 Ω resistor for the circuit shown in Figure 5.16.

81. Determine the Thévenin and Norton equivalents driving the 12 kΩ resistor for the circuit shown in Figure 5.17

82. Given the circuit of Figure 5.26, determine if the 4 kΩ resistor is the optimal value to achieve maximum power dissipation in that resistor. If it is not, determine the value that will produce maximum power in the resistor along with the resulting power.

83. For the circuit of Figure 5.36, determine an equivalent circuit using just a single voltage source.

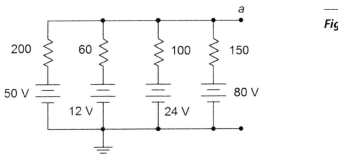

Figure 5.36

Simulation

84. Using DC bias simulations, compare the original circuit of problem 1 to its converted equivalent. Do this by connecting a resistor to the output terminals, trying several different resistance values and checking to see if the two circuits always produce the same voltage across this resistor.

85. Using DC bias simulations, compare the original circuit of problem 5 to its converted equivalent. Do this by connecting a resistor to the output terminals, trying several different resistance values and checking to see if the two circuits always produce the same voltage across this resistor.

86. Perform a DC bias simulation on the circuit of problem 11 to verify the node voltages.

87. Perform a DC bias simulation on the circuit of problem 13 to verify the node voltages.

88. Perform a DC bias simulation on the circuit of problem 19 to verify the node voltages.

89. Perform a DC bias simulation on the circuit of problem 21 to verify the resistor current.

90. Create DC bias simulations of the original and equivalent circuits generated in problem 45 to determine if the equivalent circuit is truly equivalent. Do this by substituting several different values for the 4 kΩ resistor in both circuits to see if the same load voltage is obtained for both circuits.

91. Create DC bias simulations of the original and equivalent circuits generated in problem 49 to determine if the equivalent circuit is truly equivalent. Do this by substituting several different values for the 20 kΩ resistor in both circuits to see if the same load voltage is obtained for both circuits.

92. Create DC bias simulations of the original and equivalent circuits generated in problem 53 to determine if the equivalent circuit is truly equivalent. Do this by substituting several different values for the 12 kΩ resistor in both circuits to see if the same load voltage is obtained for both circuits.

93. Perform a DC bias simulation on the circuit of problem 63 to verify that the node voltages of the new design match those of the original.

94. Perform a DC bias simulation on the circuit of problem 65 to verify that the node voltages of the new design match those of the original.

95. Using either Monte Carlo simulation or multiple DC bias simulations, verify that the resistance calculated in problem 55 achieves maximum power. Do this by trying several resistor values near the calculated value, and determining the power for each based on the squared load voltages.

96. Using either Monte Carlo simulation or multiple DC bias simulations, verify that the resistance calculated in problem 56 achieves maximum power. Do this by trying several resistor values near the calculated value, and determining the power for each based on the squared load voltages.

6 Mesh and Nodal Analysis, and Dependent Sources

This section covers:
- Series-parallel resistor circuits using multiple voltage and/or current sources via mesh and nodal analyses.
- Dependent voltage and current sources.

6.0 Introduction

Mesh Analysis

Mesh analysis uses KVL to create a series of loop equations that can be solved for mesh currents. The current through any particular component may be a mesh current or a combination of mesh currents. Circuits using complex series-parallel arrangement with multiple voltage and/or current sources may solved using this technique.

Consider the circuit of Figure 6A. We begin by designating a series of current loops. These loops should be minimal in size and cover all components at least once. By convention, the loops are drawn clockwise. There is nothing magic about them being clockwise, it is just a matter of consistency. In the circuit of Figure 6A we have two loop currents, I1 and I2. Note that all components exist in at least one loop (and sometimes in more than loop, like R3). Depending on circuit values, one or more of these loop directions may in fact be opposite of reality. This is not a problem. If this is the case, the currents will show up as negative values, and thus we know that they're really flowing counter-clockwise.

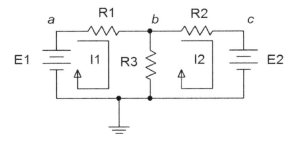

Figure 6A

We begin by writing KVL equations for each loop.

Loop 1: E1 = V of R1 + V of R3
Loop 2: −E2 = V of R2 + V of R3   (E2 is negative as I2 is drawn flowing out of its negative terminal)
Expand the voltage terms using Ohm's law.

Loop 1: E1 = I1 · R1 + (I1 − I2) R3
Loop 2: −E2 = I2 · R2 + (I2 − I1) R3

Multiplying out and collecting terms yields:

Loop 1: E1 = (R1 + R3) I1 − R3 · I2
Loop 2: −E2 = R3 · I1 + (R2 + R3) I2

As the resistor values and source voltages are known, we have two equations with two unknowns. These can be solved for I1 and I2 using simultaneous equation solution techniques such as determinants or Gauss-Jordan elimination. These equations can be obtained through inspection. Simply focus on one loop and ask the following questions: what is the total source voltage in this loop? This yields the voltage constant. Then simply sum the resistance values in the loop under inspection. This yields the coefficient for that current term. For the other current coefficients, sum the resistances that are in common between the loop under inspection and the other loops (e.g., for loop 1, R3 is in common with loop 2). These values will always be negative. As a crosscheck, the set of equations produced must exhibit *diagonal symmetry*, that is, if a diagonal is drawn from the upper left to the lower right through the IR pairs, then the coefficients found above the diagonal will have to match those found below the diagonal. In the example above, note the matching "−R3" coefficients in the final pair of equations.

While it is possible to extend this technique to include current sources, it is often easier and less error-prone to convert the current sources into voltage sources and continue with the direct inspection method outline above. In closing, it is important to remember that the number of loops determines the number of equations to be solved.

Nodal Analysis

Nodal analysis uses KCL to create a series of node equations that can be solved for node voltages. In some respects it is similar to mesh analysis. We will examine two variations; one using voltage sources, the other using current sources.

Consider the circuit shown in Figure 6B, following. We begin by labeling connection points and assigning current directions. These directions are chosen arbitrarily and may be opposite of reality. If so, their values will ultimately show up as negative.

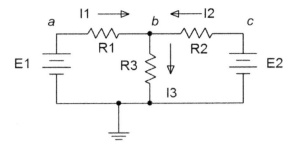

Figure 6B

Write a current summation equation for each summing node, except for ground. In this circuit there is only one node where currents combine, point b.

I1 + I2 = I3

Describe these currents in terms of the node voltages and components. For example, I3 is the node b voltage divided by R3 while I1 is the voltage across R1 divided by R1. This voltage is $V_a - V_b$.

$(V_a - V_b)/R1 + (V_c - V_b)/R2 = V_b/R3$

Noting that V_a = E1 and V_c = E2, with a little algebra this can be reduced to:

E1(1/R1) + E2(1/R2) = V_b(1/R1 + 1/R2 + 1/R3)

All quantities are known except for V_b. If there had been more nodes, there would have been an equal number of equations.

For current sources, a more direct approach is possible. Consider the circuit of Figure 6C. We start as before, identifying nodes and labeling currents. When then write current summation equations at each node (except for ground). We consider currents entering a node as positive, and exiting as negative.

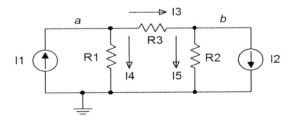

Figure 6C

81

Node a: $I1 = I3 + I4$
Node b: $I3 = I2 + I5$, and rearranging in terms of the fixed source,
Node b: $-I2 = -I3 + I5$

The currents are then described by their Ohm's law equivalents:

Node a: $I1 = (V_a - V_b)/R3 + V_a/R1$
Node b: $-I2 = -(V_a - V_b)/R3 + V_b/R5$

Expanding and collecting terms yields:

Node a: $I1 = (1/R1 + 1/R3)V_a - (1/R3)V_b$
Node b: $-I2 = -(1/R3)V_a + (1/R3 + 1/R5)V_b$

As the resistor values and currents are known, the node voltages may be solved for using simultaneous equation solution techniques. There will be as many equations as node voltages.

Like mesh analysis, there is a method to generate the equations by inspection. For the node under inspection, sum all of the current sources connected to it to obtain the current constant. The conductance term for that node will be the sum of all of the conductances connected to that node. For the other node conductances, determine the conductances between the node under inspection and these other nodes. These terms will all be negative and once again, the set of equations thus produced must exhibit diagonal symmetry (note the "$-1/R3$" coefficients for the final set of equations, above). For example, focusing on node a above, we find the fixed current source I1 feeding it (entering, therefore positive). The conductances directly connected to node a are $1/R1$ and $1/R3$, yielding the coefficient for V_a. The only conductance common between nodes a and b is $1/R3$, yielding the V_b coefficient.

Given circuits with voltage sources, it may be easier to convert them to current sources and then apply the inspection technique rather than using the general approach outlined initially. There is one trap to watch out for when using source conversions, and this also applies to mesh analysis: the voltage across or current through a converted component will most likely not be the same as the voltage or current in the original circuit. This is because the location of the converted component will have changed. For example, the circuit in Figure 6C could be solved using mesh analysis by converting the current sources and their associated resistances into current sources. That is, I1/R1 would be converted into a source E1 with a series resistor R1. Although R1 still connects to node a, the other end no longer connects to ground. Rather, it connects to the new E1. Therefore, the voltage drop across R1 in the converted circuit is not likely to equal the voltage drop seen across R1 in the original circuit (the only way they would be equal is if E1 turned out to be 0). In the converted circuit, node a has not changed from the original, so the original voltage across R1 can be determined via V_a.

Dependent Sources

A *dependent source* is a current or voltage source whose value is not fixed (i.e., independent) but rather which depends on some other circuit current or voltage. The general form for the value of a dependent source is $Y=kX$ where X and Y are currents and/or voltages and k is the proportionality factor. For example, the value of a dependent voltage source may be a function of a current, so instead of the source being equal to, say, 10 volts, it could be equal to twenty times the current passing through a particular resistor, or $V=20I$.

There are four possible dependent sources: the voltage-controlled voltage source (VCVS), the current-controlled voltage source (CCVS), the voltage-controlled current source (VCCS), and the current-controlled current source (CCCS). The source and control parameters are the same for both the VCVS and the CCCS so k is unitless (although it may be given as volts/volt and amps/amp, respectively). For the VCCS and CCVS, k has units of amps/volt and volts/amp, respectively. These are referred to as the *transresistance* and *transconductance* of the sources with units of ohms and siemens.

The schematic symbols for dependent or controlled sources are usually drawn using a diamond. Also, there will be a secondary connection for the controlling current or voltage. Examples of a voltage-controlled voltage source, current-controlled voltage source, voltage-controlled current source and a current-controlled current source are shown in Figure 6D, left-to-right. On each of these symbols, the control element is shown to the left of the source. This portion is not always drawn on a schematic. Instead, the source simply may be labeled as a function, as in $V = 0.02\ I_X$ where I_X is the controlling current.

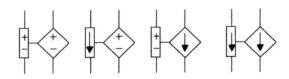

Figure 6D

Dependent sources are not "off-the-shelf" items in the same way that a battery is. Rather, dependent sources are used to model the behavior of more complex devices. For example, a bipolar junction transistor commonly is modeled as a CCCS while a field effect transistor may be modeled as a VCCS. Similarly, many op amp circuits are modeled as VCVS systems. Solutions for circuits using dependent sources follow along the lines of those established for independent sources (i.e., the application of Ohm's law, KVL, KCL, etc.), however, the sources are now dependent on the remainder of the circuit which tends to complicate the analysis. In general, there are two possible configurations: *isolated* and *coupled*. A example of the isolated form is shown in Figure 6E.

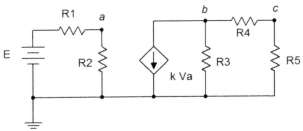

Figure 6E

In this example, the dependent source (center) does not interact with the subcircuit on the left driven by the independent source, thus it can be analyzed as two separate circuits. Solutions for this form are relatively straightforward in that the control value for the dependent source can be computed directly. The value is then substituted into the dependent source and the analysis continues as is typical. Sometimes it is convenient if the solution for a particular voltage or current is defined in terms of the control parameter rather than as a specific value (e.g., the voltage across a particular resistor might be expressed as 8 V_A instead of just 12 volts).

The second type of circuit (*coupled*) is somewhat more complex in that the dependent source can affect the parameter that controls the dependent source. In other words, the dependent source(s) will contribute terms that include the controlling parameter(s) so some additional effort will be in order. To illustrate, consider the circuit of Figure 6F.

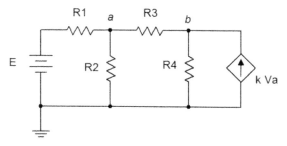

Figure 6F

In this example it should be obvious that the current from the dependent source can affect the voltage at node A, and it is this very voltage that in turn sets up the value of the current source. Circuits of this type can be analyzed using mesh or nodal analysis. Nodal analysis works well here and is illustrated below.

We begin by defining current directions. Assume that the currents through R1 and R3 are flowing into node A, the current through R2 is flowing out of node A, and the current through R4 is flowing out of node B. We shall number the branch currents to reflect the associated resistor. The resulting KCL equations are:

Node a: $I1 + I3 = I2$
Node b: $k V_a = I3 + I4$

The currents are then described by their Ohm's law equivalents:

Node a: $(E - V_a)/R1 + (V_b - V_a)/R3 = V_a/R2$
Node b: $k V_a = (V_b - V_a)/R2 + V_b/R4$

Expanding terms yields:

Node a: $E/R1 - V_a/R1 + V_b/R3 - V_a/R3 = V_a/R2$
Node b: $k V_a = V_b/R2 - V_a/R2 + V_b/R4$

Collecting terms and simplifying yields:

Node a: $E/R1 = (1/R1 + 1/R2 + 1/R3) V_a - (1/R3) V_b$
Node b: $0 = -(k + 1/R2) V_a + (1/R2 + 1/R4) V_b$

Values for the resistors, k and E are known, so the analysis proceeds as usual.

Also, it worth remembering that it is possible to perform source conversions on dependent sources, within limits. The same procedure is followed as for independent sources. The new source will remain a dependent source (e.g., converting VCVS to VCCS). This process is not applicable if the control parameter directly involves the internal impedance (i.e., is its voltage or current).

Notes

6.1 Exercises

Analysis

1. Given the circuit in Figure 6.1, write the mesh loop equations.

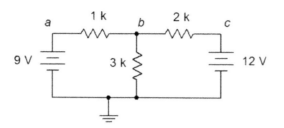
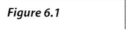

Figure 6.1

2. Using mesh analysis, determine the value of V_b for the circuit shown in Figure 6.1.
3. For the circuit shown in Figure 6.1, use mesh analysis to determine the current through the 1 kΩ resistor.
4. Given the circuit in Figure 6.2, write the mesh loop equations and the associated determinants.

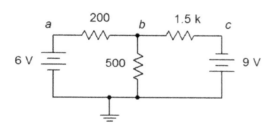

Figure 6.2

5. Using mesh analysis, determine the value of V_b for the circuit shown in Figure 6.2.
6. For the circuit shown in Figure 6.2, use mesh analysis to determine the current through the 500 Ω resistor.

7. Given the circuit in Figure 6.3, write the mesh loop equations.

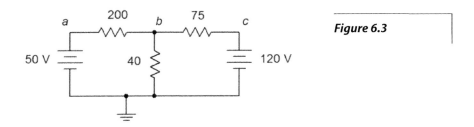

Figure 6.3

8. Using mesh analysis, determine the value of V_b for the circuit shown in Figure 6.3.

9. For the circuit shown in Figure 6.3, use mesh analysis to determine the current through the 75 Ω resistor.

10. Given the circuit in Figure 6.4, write the mesh loop equations and the associated determinants.

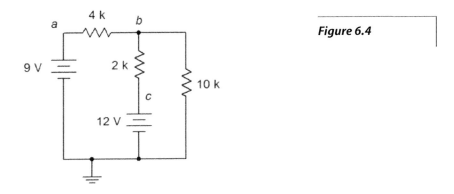

Figure 6.4

11. Using mesh analysis, determine the value of V_{bc} for the circuit shown in Figure 6.4.

12. For the circuit shown in Figure 6.4, use mesh analysis to determine the current through the 10 kΩ resistor.

13. Given the circuit in Figure 6.5, write the mesh loop equations.

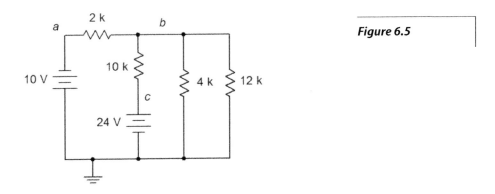

Figure 6.5

14. Using mesh analysis, determine the value of V_{ac} for the circuit shown in Figure 6.5.

15. For the circuit shown in Figure 6.5, use mesh analysis to determine the current through the 4 kΩ resistor.

16. Given the circuit in Figure 6.6, write the mesh loop equations and the associated determinants.

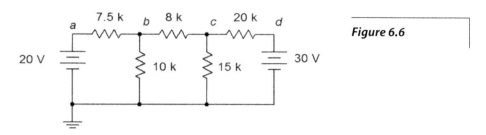

Figure 6.6

17. Using mesh analysis, determine the value of V_c for the circuit shown in Figure 6.6.

18. For the circuit shown in Figure 6.6, use mesh analysis to determine the current through the 10 kΩ resistor.

19. Given the circuit in Figure 6.7, write the mesh loop equations.

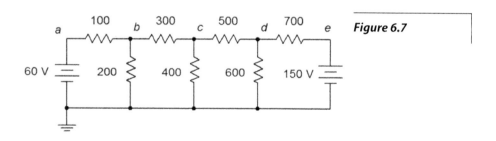

Figure 6.7

20. Using mesh analysis, determine the value of V_{bd} for the circuit shown in Figure 6.7.

21. For the circuit shown in Figure 6.7, use mesh analysis to determine the current through the 500 Ω resistor.

22. Given the circuit in Figure 6.8, write the mesh loop equations.

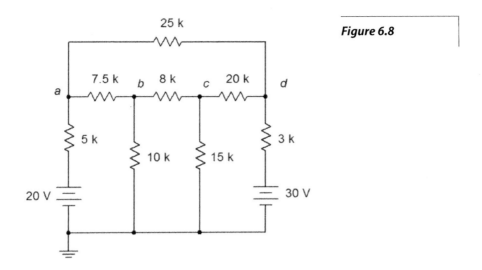

Figure 6.8

23. Using mesh analysis, determine the value of V_{ad} for the circuit shown in Figure 6.8.

24. For the circuit shown in Figure 6.8, use mesh analysis to determine the current through the 3 kΩ resistor.

25. Given the circuit in Figure 6.9, write the mesh loop equations.

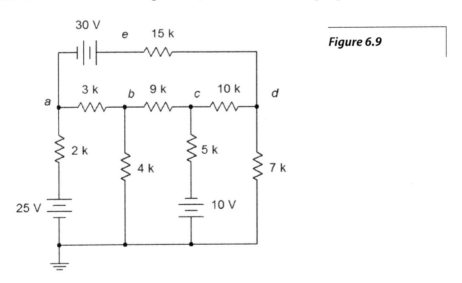

Figure 6.9

26. Using mesh analysis, determine the value of V_e for the circuit shown in Figure 6.9.

27. For the circuit shown in Figure 6.9, use mesh analysis to determine the current through the 9 kΩ resistor.

28. Given the circuit in Figure 6.10, write the mesh loop equations and the associated determinants.

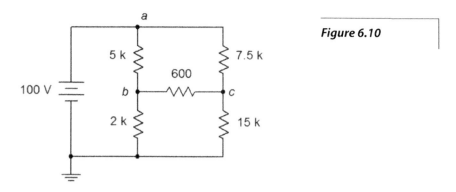

Figure 6.10

29. Using mesh analysis, determine the value of V_{bc} for the circuit shown in Figure 6.10.

30. Given the circuit shown in Figure 6.10, use mesh analysis to determine the current through the 600 Ω resistor.

31. Given the circuit in Figure 6.11, write the mesh loop equations.

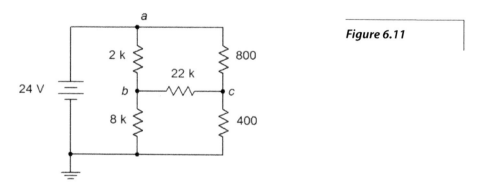

Figure 6.11

32. Using mesh analysis, determine the value of V_{bc} for the circuit shown in Figure 6.11.

33. Given the circuit shown in Figure 6.11, use mesh analysis to determine the current through the 800 Ω resistor.

34. For the circuit in Figure 6.12, write the mesh loop equations.

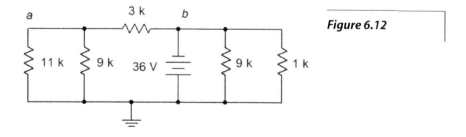

Figure 6.12

90

35. Using mesh analysis, determine the value of V_a for the circuit shown in Figure 6.12.

36. For the circuit shown in Figure 6.12, use mesh analysis to determine the current through the 3 kΩ resistor.

37. Given the circuit in Figure 6.13, write the mesh loop equations.

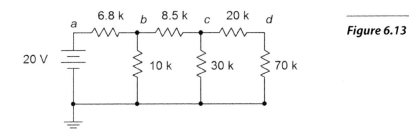

Figure 6.13

38. Using mesh analysis, determine the value of V_c for the circuit shown in Figure 6.13.

39. For the circuit shown in Figure 6.13, use mesh analysis to determine the current passing through the 8.5 kΩ resistor.

40. Given the circuit in Figure 6.14, write the mesh loop equations (consider using source conversion).

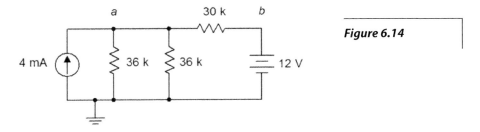

Figure 6.14

41. Using mesh analysis, determine the value of V_a for the circuit shown in Figure 6.14.

42. For the circuit shown in Figure 6.14, use mesh analysis to determine the current passing through the 30 kΩ resistor.

43. Given the circuit in Figure 6.15, write the mesh loop equations (consider using source conversion).

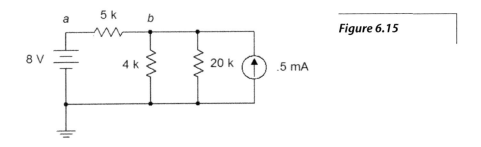

Figure 6.15

44. Using mesh analysis, determine the value of V_b for the circuit shown in Figure 6.15.

45. For the circuit shown in Figure 6.15, use mesh analysis to determine the current through the 5 kΩ resistor.

46. Given the circuit in Figure 6.16, write the node equations.

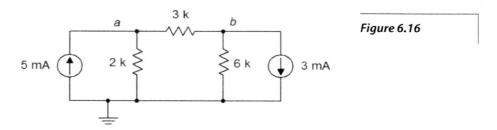

Figure 6.16

47. Using nodal analysis, determine the value of V_b for the circuit shown in Figure 6.16.

48. For the circuit shown in Figure 6.16, use nodal analysis to determine the current through the 3 kΩ resistor.

49. Given the circuit in Figure 6.17, write the node equations.

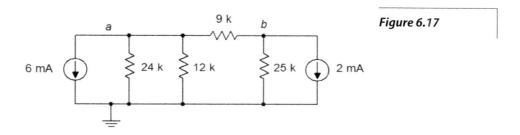

Figure 6.17

50. Using nodal analysis, determine the value of V_b for the circuit shown in Figure 6.17.

51. For the circuit shown in Figure 6.17, use nodal analysis to determine the current passing through the 12 kΩ resistor.

52. Given the circuit in Figure 6.18, write the node equations.

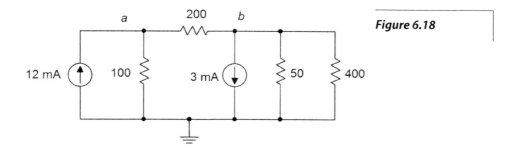

Figure 6.18

53. Using nodal analysis, determine the value of V_{ba} for the circuit shown in Figure 6.18.

54. For the circuit shown in Figure 6.18, use nodal analysis to determine the current passing through the 100 Ω resistor.

55. Given the circuit in Figure 6.19, write the node equations.

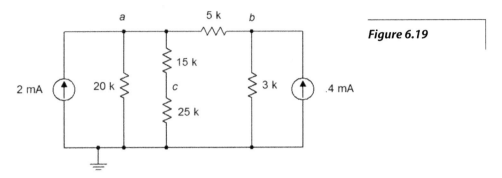

Figure 6.19

56. Using nodal analysis, determine the value of V_{ac} for the circuit shown in Figure 6.19.

57. For the circuit shown in Figure 6.19, use nodal analysis to determine the current passing through the 20 kΩ resistor.

58. Given the circuit in Figure 6.20, write the node equations.

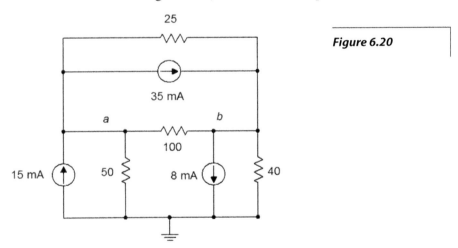

Figure 6.20

59. Using nodal analysis, determine the value of V_b for the circuit shown in Figure 6.20.

60. For the circuit shown in Figure 6.20, use nodal analysis to determine the current through the 3 kΩ resistor.

61. Given the circuit in Figure 6.21, write the node equations.

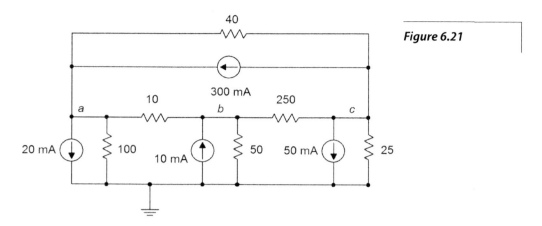

Figure 6.21

62. Using nodal analysis, determine the value of V_b for the circuit shown in Figure 6.21.

63. For the circuit shown in Figure 6.21, use nodal analysis to determine the current through the 40 Ω resistor.

64. Given the circuit in Figure 6.13, write the node equations.

65. Using nodal analysis, determine the value of V_d for the circuit shown in Figure 6.13.

66. For the circuit shown in Figure 6.13, use nodal analysis to determine the current passing through the 20 kΩ resistor.

67. Given the circuit in Figure 6.14, write the node equations using the general approach. Do not use source conversions.

68. Using nodal analysis, determine the value of V_{ab} for the circuit shown in Figure 6.14.

69. For the circuit shown in Figure 6.14, use nodal analysis to determine the current passing through the 30 kΩ resistor.

70. Given the circuit in Figure 6.15, write the node equations.

71. Using nodal analysis, determine the value of V_b for the circuit shown in Figure 6.15.

72. For the circuit shown in Figure 6.15, use nodal analysis to determine the current through the 4 kΩ resistor.

73. For the circuit of Figure 6.22, determine V_b.

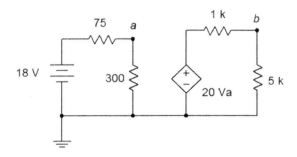

Figure 6.22

74. For the circuit of Figure 6.23, determine V_c.

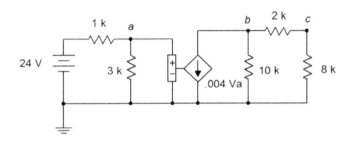

Figure 6.23

75. Given the circuit of Figure 6.24, determine V_b.

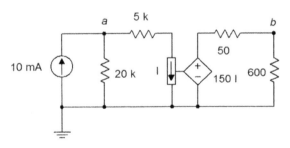

Figure 6.24

76. Find the current through the 10 kΩ resistor given the circuit of Figure 6.25.

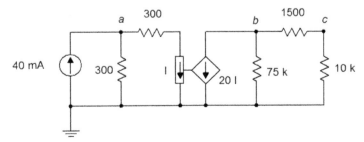

Figure 6.25

77. Given the circuit of Figure 6.26, determine V_c.

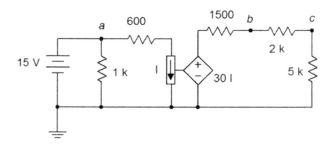

Figure 6.26

78. In the circuit of Figure 6.27, determine V_a.

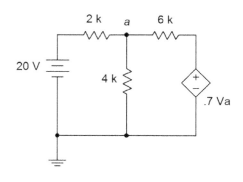

Figure 6.27

79. For the circuit of Figure 6.28, determine V_a.

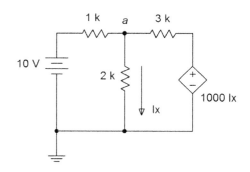

Figure 6.28

80. For the circuit of Figure 6.29, determine V_a.

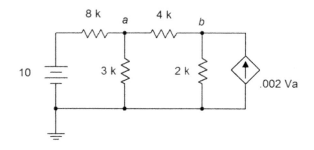

Figure 6.29

Challenge

81. Given the circuit in Figure 6.8, write the node equations.

82. Using nodal analysis, determine the value of V_b for the circuit shown in Figure 6.8.

83. For the circuit shown in Figure 6.8, use nodal analysis to determine the current through the 3 kΩ resistor.

84. Given the circuit in Figure 6.6, write the node equations.

85. Using nodal analysis, determine the value of V_c for the circuit shown in Figure 6.6.

86. For the circuit shown in Figure 6.6, use nodal analysis to determine the current through the 8 kΩ resistor.

87. Given the circuit in Figure 6.10, write the node equations and the associated determinants.

88. Using nodal analysis, determine the value of V_{bc} for the circuit shown in Figure 6.10.

89. For the circuit shown in Figure 6.10, use nodal analysis to determine the current through the 2 kΩ resistor.

90. Given the circuit of Figure 6.30, determine V_c.

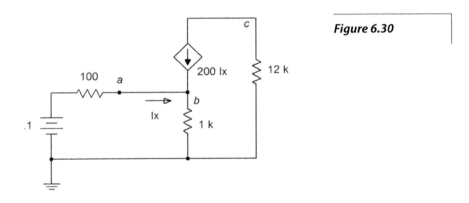

Figure 6.30

91. Find the current through the 10 kΩ resistor in the circuit of Figure 6.31.

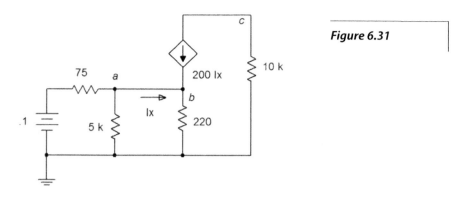

Figure 6.31

92. Given the circuit of Figure 6.32, determine V_c.

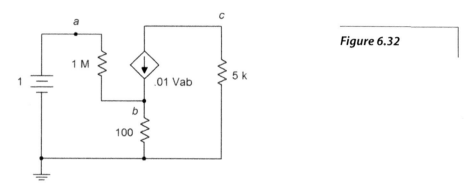

Figure 6.32

93. Given the circuit of Figure 6.33, determine the current through the 5 kΩ resistor.

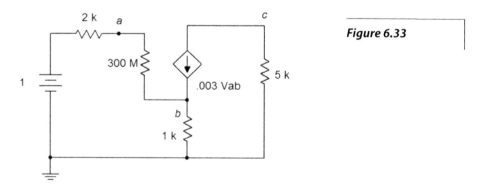

Figure 6.33

94. For the circuit of Figure 6.34, determine V_b.

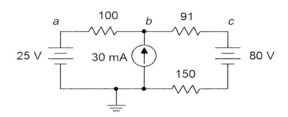

Figure 6.34

95. For the circuit of Figure 6.35, determine V_c.

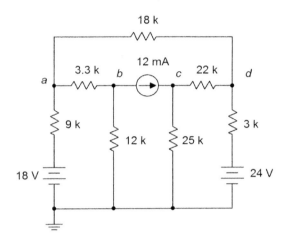

Figure 6.35

96. For the circuit of Figure 6.36, determine V_a.

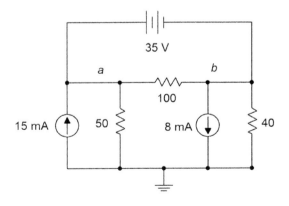

Figure 6.36

97. For the circuit of Figure 6.37, determine V_b.

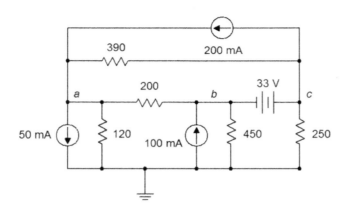

Figure 6.37

Simulation

98. Perform a DC bias simulation on the circuit depicted in Figure 6.7 to verify the component currents.

99. Perform a DC bias simulation on the circuit depicted in Figure 6.9 to verify the component currents.

100. Perform a DC bias simulation on the circuit depicted in Figure 6.7 to verify the loop currents and node voltages.

101. Perform a DC bias simulation on the circuit depicted in Figure 6.16 to verify the node voltages.

102. Perform a DC bias simulation on the circuit depicted in Figure 6.19 to verify the node voltages.

103. Perform a DC bias simulation on the circuit depicted in Figure 6.20 to verify the node voltages.

7 Inductors and Capacitors

This section covers:
- Definitions for inductors and capacitors.
- Initial and steady-state analysis of RLC circuits.
- Transient response of RC and RL circuits.

7.0 Introduction

Inductors and Capacitors

Inductors and capacitors are both devices that store energy. In the case of the inductor, energy is stored in a magnetic field and in the case of the capacitor, an electric field. Unlike a battery, the energy storage times are typically very short, perhaps only milliseconds or microseconds.

A capacitor consists of two conducting plates separated by an insulator, or *dielectric*. This material can be air or made from a variety of different materials such as plastics and ceramics. In general, capacitance increases directly with plate area, A, and inversely with plate separation distance, d. Further, it is also proportional to the permitivity, ε, of the dielectric. Thus, capacitance in farads is equal to:

$$C = \varepsilon \frac{A}{d}$$

The breakdown strength of the dielectric will set an upper limit on how large of a voltage may be placed across a capacitor before it is damaged. Breakdown strength is measured in volts per unit distance, thus, the closer the plates, the less voltage the capacitor can withstand. Thus, halving the plate distance doubles the capacitance but also halves its voltage rating.

Placing capacitors in parallel increases overall plate area, and thus increases capacitance. Therefore capacitors in parallel add in value. When placed in series, it is as if the plate distance has increased, thus decreasing capacitance. Therefore capacitors in series behave like resistors in parallel and the equivalent value can be found via summing reciprocal values or via the product-sum rule.

The fundamental current-voltage relationship of a capacitor is not the same as that of resistors. The current through a capacitor is equal to the capacitance times the rate of change of the capacitor voltage. That is, the value of the voltage is not important, but rather how quickly the voltage is changing. Given a fixed voltage, the capacitor current is zero and thus the capacitor behaves like an open. If the voltage is changing rapidly, the current will be high and the capacitor behaves more like a short.

Expressed as a formula:

$$i = C \frac{dv}{dt}$$

Note that if a capacitor is driven by a fixed current source, the voltage across it rises at the constant rate of i/C. Also, the voltage across the capacitor cannot change instantaneously. This would mean that dv/dt would be infinite, and thus, the current driving the capacitor would also have to be infinite (an impossibility).

The charge developed across a capacitor is equal to the voltage times the capacitance, or $Q = v \cdot C$. A slight rearrangement yields $C = Q/v$, showing that the larger the capacitor, the more charge it can hold per volt impressed across it.

If multiple capacitors are in series across a voltage source, as shown in Figure 7A, the steady state voltage will divide between them in inverse proportion. The voltages can also be found by determining the series equivalent capacitance, and using the applied voltage, determining the total charge. The individual voltages can then be computed from $v = Q/C$, where Q is the total charge and C is the capacitance of interest.

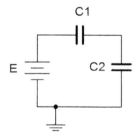

Figure 7A

An inductor in its simplest form consists of a series of wire loops. These might be wound around an iron core, although a non ferrous core might also be used. For a simple single layer inductor, the inductance is proportional to the permeability, μ, of the core material; the cross sectional area of the loops, A; and the square of the number of loops, N. It is inversely proportional to the length, l. As a formula, the inductance in henries is:

$$L = \mu \frac{A N^2}{l}$$

Thus, we can see that doubling the number of loops would double the inductance. This is because although N^2 goes up by a factor of four, the length would also have to double in order to accommodate these extra loops. Consequently, inductors in series add values just like resistors.

The fundamental current-voltage relationship of the inductor is the mirror image of the capacitor.

$$v = L \frac{di}{dt}$$

This states that the voltage across the inductor is a function of how quickly the current is changing. If the current is not changing (i.e., in steady state), then the voltage across the inductor is zero. In this case, the inductor behaves like a short. In contrast, during a rapid initial current change, the inductor voltage can be large, and thus the inductor behaves like an open.

Initial and Steady-State Analysis of RLC Circuits

When analyzing resistor-inductor-capacitor circuits, remember that capacitor voltage cannot change instantaneously, thus, initially, capacitors behave as a short circuit. Once the capacitor has been charged and is in a steady state condition, it behaves like an open. The inductor behaves in the opposite manner. Current through an inductor cannot change instantaneously as this would require an infinite voltage source. Thus, initially an inductor behaves like an open, but once steady state is reached, it behaves like a short. For example, in the circuit of Figure 7B, initially L is open and C is a short, leaving us with R1 and R2 in series with the source, E. At steady state, L shorts out both C and R2, leaving all of E to drop across R1. All practical inductors will exhibit some internal resistance, so it is often best to think of an inductor as an ideal inductance with a small resistance in series with it. Similarly, practical capacitors can be thought of as an ideal capacitance in parallel with a very large (leakage) resistance.

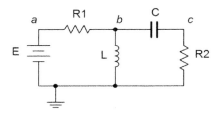

Figure 7B

Transient Response of RC Circuits

The question remains, "What happens between the time the circuit is powered up and when it reached steady state?" This is known as the *transient response*. Consider the circuit shown in Figure 7C.

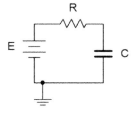

Figure 7C

The key to the analysis is to remember that capacitor voltage cannot change instantaneously. Once power is applied, the capacitor voltage must be zero, and all of the source voltage drops across the resistor. This creates the initial current, and this current starts to charge the capacitor (the initial rate being equal to i/C). According to Kirchhoff's voltage law, as the capacitor voltage begins to increase, the resistor voltage must decrease because the sum of the two must equal the fixed source voltage. This means that the circulating current must also decrease. This, in turn, means that the rate of capacitor voltage increase begins to slow. As the capacitor voltage continues to increase, less voltage is available for the resistor, causing further reductions in current, and a further slowing of the rate of capacitor voltage change. Eventually, the capacitor voltage will be nearly equal to the source voltage. This will result in a very small potential across the resistor and an equally small current, slowing subsequent capacitor voltage increases to a near standstill. Theoretically, the capacitor voltage approaches the source voltage but never quite equals it. Similarly, the current drops to near zero, but never completely turns off. This is illustrated in Figure 7D.

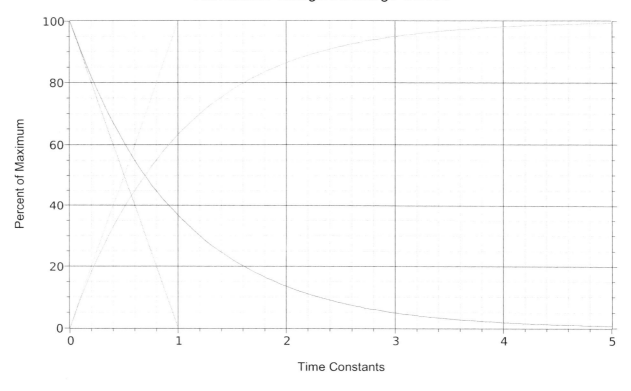

Figure 7D

The dotted red line represents the initial rate of change of capacitor voltage. This trajectory is what would be expected if an ideal current source drove the capacitor. As noted previously, the rate is equal to i/C, and therefore in this case, E/RC. If the initial rate of change were to continue unabated, the source voltage, E, would be reached in RC seconds. Consequently, RC is referred to as the charge time constant and is denoted by τ (Greek letter tau).
Thus,

$$\tau = RC$$

As noted, once the capacitor begins to charge, the current begins to decrease and the capacitor voltage curve begins to fall away from the initial trajectory. The solid red curve represents the capacitor voltage. *After five time constants the capacitor is nearly fully charged and the circuit is considered to be in steady state* (i.e., the capacitor behaves as an open). The equation for this curve is:

$$v_C = E\left(1 - \epsilon^{\frac{-t}{\tau}}\right)$$

The dotted blue line shows the initial slope of current change. The solid blue curve shows the circulating current (and the resistor voltage). The equation for this curve is:

$$i = \frac{E}{R}\epsilon^{\frac{-t}{\tau}} \quad \text{and}$$

$$v_R = E\epsilon^{\frac{-t}{\tau}}$$

If a more complex circuit is used, the section feeding the capacitor may be simplified using Thévenin's Theorem to determine the effective source voltage and charging resistance.

If power is interrupted before the capacitor is fully charged, the equations above may be used to determine the precise voltage(s) and current reached. The capacitor will then behave as a voltage source and begin to discharge, its voltage curve following the blue plot line, above. It is possible for the discharge circuit to have a different resistance than the charging circuit, and thus, a different time constant. Finally, because the capacitor is discharging, its current direction will be opposite to that of the charging current. KVL must still be satisfied, but because the capacitor is now behaving as a source, the magnitude of the discharge resistance's voltage must equal the capacitor voltage magnitude. Therefore, its curve will take the same shape (the solid blue curve).

Transient Response of RL Circuits

The transient response of RL circuits is nearly the mirror image of RC circuits. Consider the circuit of Figure 7E.

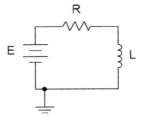

Figure 7E

The key to this analysis is to remember that inductor current cannot change instantaneously. When power is first applied, the circulating current must remain at zero. Therefore no voltage drop is produced across the resistor, and by KVL, the voltage across the inductor must equal the source, E. This establishes the initial rate of change of current (E/L) and is represented by the dotted red line in the graph of Figure 7D. As the current starts to increase, the voltage drop across the resistor begins to increase. This reduces the voltage available for the inductor, thus slowing the rate of change of current. This is depicted by the solid red curve on the graph. Meanwhile, the solid blue curve represents the decreasing inductor voltage. Thus, in the RL circuit, the inductor's voltage curve echoes the RC circuit's current curve (or resistor voltage curve), and the RL current curve echoes the RC circuit's capacitor voltage curve. The time constant for an RL circuit is:

$$\tau = \frac{L}{R}$$

Once again, five constants will achieve steady state. The relevant equations are:

$$v_L = E \epsilon^{\frac{-t}{\tau}}$$

$$i = \frac{E}{R}\left(1 - \epsilon^{\frac{-t}{\tau}}\right) \quad \text{and}$$

$$v_R = E\left(1 - \epsilon^{\frac{-t}{\tau}}\right)$$

As in the RC case, complex circuits may need to be simplified first using Thévenin's Theorem to determine the effective source voltage and charging resistance. Again, the discharge resistance may be considerably different from the charging resistance. Finally, if an RL circuit is abruptly altered or opened, very large voltage spikes may occur. This is due to the fact that inductor current cannot change instantaneously. Thus, if the circuit is opened, the open represents a very large resistance. Ohm's law indicates that the inductor current times this very large resistance may produce a very large voltage across the new open. In fact, the potential may be sufficient to cause a spark or arc. Note that because the current cannot change instantaneously (both magnitude and direction), the inductor now behaves as a voltage source of very high magnitude and with reverse polarity.

7.1 Exercises

Analysis

1. For the circuit shown in Figure 7.1, determine the effective capacitance.

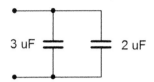
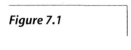

Figure 7.1

2. Determine the effective capacitance of the configuration shown in Figure 7.2.

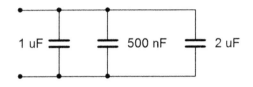

Figure 7.2

3. Given the capacitor network shown in Figure 7.3, determine the effective value.

Figure 7.3

4. Determine the effective capacitance of network shown in Figure 7.4.

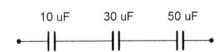

Figure 7.4

5. For the circuit shown in Figure 7.5, determine the effective inductance.

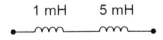
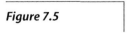

Figure 7.5

6. Determine the effective inductance of the configuration shown in Figure 7.6.

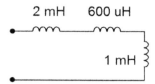

Figure 7.6

7. Given the inductor network shown in Figure 7.7, determine the effective value.

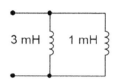

Figure 7.7

8. Determine the effective inductance of network shown in Figure 7.8.

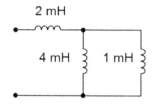

Figure 7.8

9. Determine the voltage across each capacitor for the circuit shown in Figure 7.9.

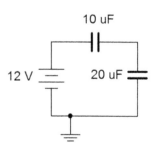

Figure 7.9

108

10. Determine the voltage across each capacitor for the circuit shown in Figure 7.10.

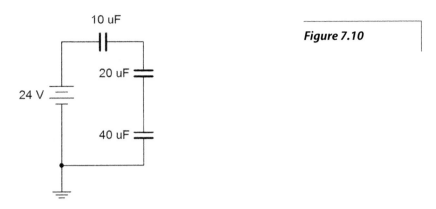

Figure 7.10

11. Determine the initial voltage across each component for the circuit shown in Figure 7.11.

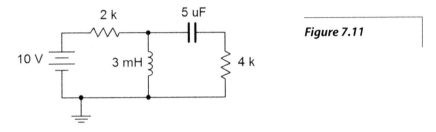

Figure 7.11

12. Given the network shown in Figure 7.11, determine the steady state voltage across each component.

13. Given the network shown in Figure 7.12, determine the steady state voltage across each component.

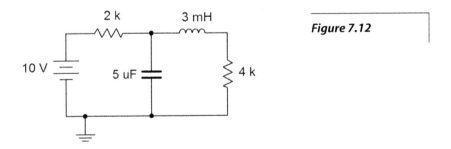

Figure 7.12

14. Determine the initial voltage across each component for the circuit shown in Figure 7.12.

15. Determine the initial voltage across each component for the circuit shown in Figure 7.13.

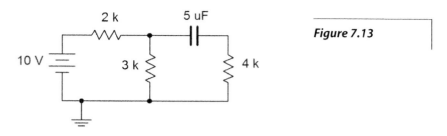
Figure 7.13

16. Given the network shown in Figure 7.13, determine the steady state voltage across each component.

17. Given the network shown in Figure 7.14, determine the steady state voltage across each component.

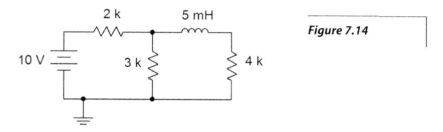
Figure 7.14

18. Determine the initial voltage across each component for the circuit shown in Figure 7.14.

19. Determine the initial voltage across each component for the circuit shown in Figure 7.15.

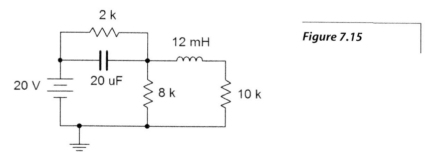
Figure 7.15

20. Given the network shown in Figure 7.15, determine the steady state voltage across each component.

21. Given the network shown in Figure 7.16, determine the steady state voltage across each component.

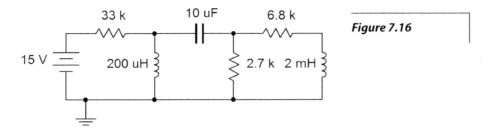

Figure 7.16

22. Determine the initial voltage across each component for the circuit shown in Figure 7.16.

23. For the circuit shown in Figure 7.17, determine the capacitor voltage 3 microseconds after the power is switched on.

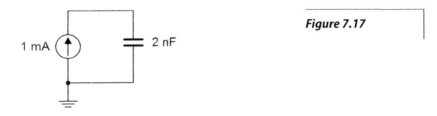

Figure 7.17

24. For the circuit shown in Figure 7.18, determine the capacitor voltage 5 seconds after the power is switched on.

Figure 7.18

25. For the circuit shown in Figure 7.19, determine the inductor current 20 microseconds after the power is switched on. Assume this is an ideal inductor with no internal resistance.

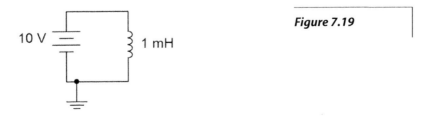

Figure 7.19

26. For the circuit shown in Figure 7.20, determine the inductor current 100 milliseconds after the power is switched on. Assume this is an ideal inductor with no internal resistance.

Figure 7.20

27. Determine the time constant and the time required to reach steady state for the circuit shown in Figure 7.21.

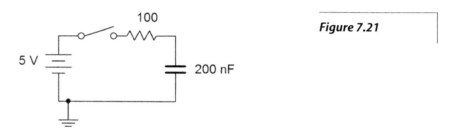

Figure 7.21

28. For the circuit shown in Figure 7.21, determine the capacitor voltage and circulating current 20 microseconds and 100 milliseconds after the switch is thrown.

29. Given the circuit shown in Figure 7.22, determine the capacitor voltage and circulating current 200 milliseconds and 10 seconds after the switch is thrown.

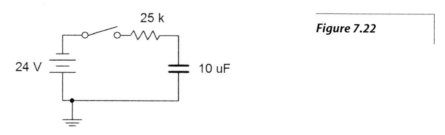

Figure 7.22

30. Determine the time constant and the time required to reach steady state for the circuit shown in Figure 7.22.

31. Determine the time constant and the time required to reach steady state for the circuit shown in Figure 7.23, switch position 1.

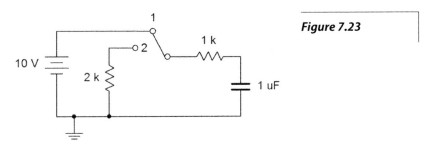

Figure 7.23

32. Determine the charge and discharge time constants for the circuit shown in Figure 7.24.

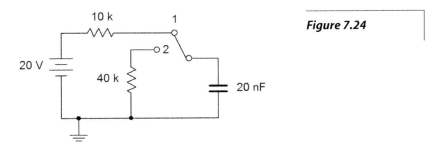

Figure 7.24

33. Given the circuit shown in Figure 7.23, determine the capacitor voltage 10 milliseconds after the power is turned on. At this point, the switch is thrown to position 2. Determine how long it will take the capacitor to discharge to nearly zero volts.

34. For the circuit shown in Figure 7.24, determine the capacitor voltage 400 microseconds after the power is turned on. At this point, the switch is thrown to position 2. Determine how long it will take the capacitor to discharge to nearly zero volts.

35. Determine the time constant and the time required to reach steady state for the circuit shown in Figure 7.25, switch position 1.

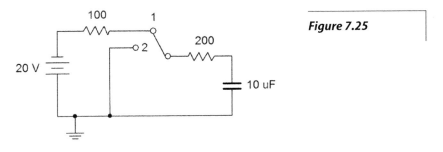

Figure 7.25

36. Given the circuit shown in Figure 7.25, determine the capacitor voltage 12 milliseconds after the power is turned on. At this point, the switch is thrown to position 2. Determine how long it will take the capacitor to discharge to nearly zero volts.

37. Determine the time constant and the time required to reach steady state for the circuit shown in Figure 7.26.

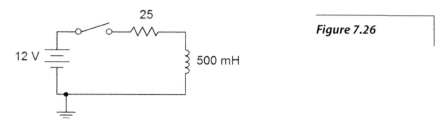

Figure 7.26

38. Determine the time constant and the time required to reach steady state for the circuit shown in Figure 7.27.

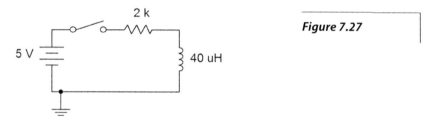

Figure 7.27

39. Given the circuit shown in Figure 7.26, determine the inductor voltage and circulating current 200 milliseconds and 2 seconds after the switch is thrown. How does this change if we include 5 Ω of internal resistance to the inductor?

40. For the circuit shown in Figure 7.27, determine the inductor voltage and circulating current 50 nanoseconds and 100 milliseconds after the switch is thrown.

41. Determine the time constant and the time required to reach steady state for the circuit shown in Figure 7.28, switch position 1.

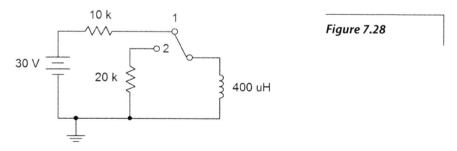

Figure 7.28

42. Determine the charge and discharge time constants for the circuit shown in Figure 7.29.

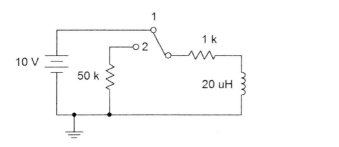
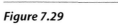

Figure 7.29

43. Given the circuit shown in Figure 7.28, determine the inductor current 1 millisecond after the power is turned on. At this point, the switch is thrown to position 2. Determine the inductor voltage the instant of switch contact (assume ideal switch).

44. For the circuit shown in Figure 7.29, determine the inductor current 400 microseconds after the power is turned on. At this point, the switch is thrown to position 2. Determine the inductor voltage the instant of switch contact (assume ideal switch).

45. Determine the time constant and the time required to reach steady state for the circuit shown in Figure 7.30, switch position 1.

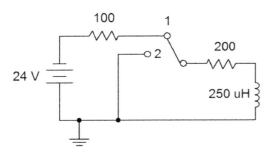

Figure 7.30

46. Given the circuit shown in Figure 7.30, determine both the inductor current and voltage 10 milliseconds after the power is turned on. At this point, the switch is thrown to position 2. Determine how long it will take the inductor to discharge to nearly zero amps. Assume ideal switch.

Design

47. Given the circuit of Figure 7.21, determine a new resistor value such that steady state is reached in 2 milliseconds.

48. Given the circuit of Figure 7.22, determine a new resistor value such that steady state is reached in 5 seconds.

49. When an audio amplifier is turned on, the power surge can cause an audible pop from the loudspeaker. To prevent this, amplifiers often connect to the loudspeaker via a relay. The relay is energized to connect the loudspeaker once the output has settled, typically a few seconds after power is applied. This delay may be created via an RC network. Suppose the driving circuit is 5 volts and the relay trips at 4 volts. Further, the associated charging resistance is 10 kΩ. Determine the capacitance required to achieve a 2 second delay time.

Challenge

50. Determine the time constant and the time required to reach steady state for the circuit shown in Figure 7.31.

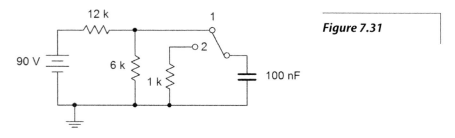

Figure 7.31

51. For the circuit shown in Figure 7.31, determine the capacitor voltage 1 second after the power is turned on. At this point, the switch is thrown to position 2. Determine how long it will take the capacitor to discharge to nearly zero volts.

52. Determine the time constant and the time required to reach steady state for the circuit shown in Figure 7.32.

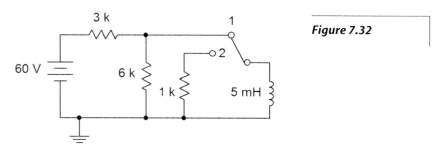

Figure 7.32

53. For the circuit shown in Figure 7.32, determine the inductor voltage 400 microseconds after the power is turned on. At this point, the switch is thrown to position 2. Determine how long it will take the inductor to discharge to nearly zero amps. Assume ideal switch.

116

Simulation

54. Perform a transient analysis to verify the time to steady state of Figure 7.21 (problem 27).

55. Perform a transient analysis to verify the time to steady state of Figure 7.22 (problem 30).

56. Use a transient analysis to verify the time constant and time to steady state of Figure 7.27 (problem 38).

57. Use a transient analysis to verify the inductor voltage waveform of Figure 7.26 as specified in problem 39.

58. Use a transient analysis to verify the design of problem 47.

59. Use a transient analysis to verify the design of problem 48.

60. Use a transient analysis to verify the operation of the circuit shown in Figure 7.31 as specified in problem 51. You may wish to do this as two separate simulations, one for each switch position, with the second position using a pre-charged capacitor.

Appendix A

Standard Component Sizes

Passive components (resistors, capacitors and inductors) are available in standard sizes. The tables below are for resistors. The same digits are used in subsequent decades up to at least 1 Meg ohm (higher decades are not shown). Capacitors and inductors are generally not available in as many standard values as are resistors. Capacitors below 10 nF (.01 µF) are usually available at the 5% standard digits while larger capacitances tend to be available at the 20% standards.

5% and 10% standard values, EIA E24 and EIA E12

10% values (EIA E12) are **bold**
20% values (seldom used) are every fourth value starting from 10
(i.e., every other 10% value)

10	11	**12**	13	**15**	16	**18**	20	**22**	24	**27**	30
33	36	**39**	43	**47**	51	**56**	62	**68**	75	**82**	91

1% and 2% standard values, EIA E96 and EIA E48

2% values (EIA E48) are **bold**

10.0	10.2	**10.5**	10.7	**11.0**	11.3	**11.5**	11.8	**12.1**	12.4	**12.7**	13.0
13.3	13.7	**14.0**	14.3	**14.7**	15.0	**15.4**	15.8	**16.2**	16.5	**16.9**	17.4
17.8	18.2	**18.7**	19.1	**19.6**	20.0	**20.5**	21.0	**21.5**	22.1	**22.6**	23.2
23.7	24.3	**24.9**	25.5	**26.1**	26.7	**27.4**	28.0	**28.7**	29.4	**30.1**	30.9
31.6	32.4	**33.2**	34.0	**34.8**	35.7	**36.5**	37.4	**38.3**	39.2	**40.2**	41.2
42.2	43.2	**44.2**	45.3	**46.4**	47.5	**48.7**	49.9	**51.1**	52.3	**53.6**	54.9
56.2	57.6	**59.0**	60.4	**61.9**	63.4	**64.9**	66.5	**68.1**	69.8	**71.5**	73.2
75.0	76.8	**78.7**	80.6	**82.5**	84.5	**86.6**	88.7	**90.9**	93.1	**95.3**	97.6

Appendix B

Answers to Selected Numbered Problems

1 Fundamentals

1. 14.54, 30060, 76.90, 0.0008475
3. 2.361E1, 1.2E4, 7.632E3, 5.09E−3
5. 12E3 (or 12 k), 470, 6.5, 1.98E−3 (or 1.98 m)
7. 1.602E−7 coulombs
9. 1.248E20 electrons
11. 0.4 amps
13. 0.5 coulombs
15. 0.2 volts
17. 20 joules
19. 1492 watts
21. 24 watts
23. 82.9%
25. 125 watts
27. 3 Ah
29. 1200 hours
31. 160 ohms
33. 8 ohms
35. 1200 ohms
37. 3.2 ohm-cm
39. yes, +2.26%
41. yes, −0.2%
43. 200 k, 68 k, 2.7 k
45. 3.3 M, 91, 390
47. 160 k to 240 k, 54.4 k to 81.6 k, 2.43 k to 2.97 k
49. 2.97 M to 3.63 M, 81.9 to 100.1, 370.5 to 409.5
51. yellow-violet-black-gold, red-red-orange-gold, orange-white-yellow-gold, brown-red-orange-gold, violet-green-brown-gold.

2 Series Resistive Circuits

1. 120 mA

3. 1.44 W

5. 15 V

7. 33 V

9. 2.6 k

11. 40 mA

13. 320 mW (200), 160 mW (100), 480 mW (source)

15. 6 V (2 k), 12 V (4 k), −6 V

17. 0.25 mA

19. 15 mA, +/− left to right on 200, +/− top to bottom on 100, +/− right to left on 500, +/− top to bottom on source.

21. 180 mW

23. 2.5 V (50), 0.5 V (10), 3 V (60), V_c = 3 V, V_{ac} = 3 V, V_a = 6 V

25. 200 mA, +/− left to right on 25, +/− top to bottom on 10, +/− right to left on 5, +/− top to bottom on source.

27. 2 V (400), 1 V (200), V_b = 10 V, V_c = 1 V, V_{ac} = 11 V

29. 3.6 V (2 k), 14.4 V (8 k), V_b = 9.6 V, V_c = -14.4 V, V_{ac} = 20.4 V

31. 35 mA, +/− left to right on 400, +/− right to left on 200, +/− top to bottom on 12 V source, +/− bottom to top on 9 V source.

33. 1.8 mA, +/− left to right on 2 k, +/− right to left on 8 k, +/− top to bottom on sources.

35. V_b = 14.4 V, V_c = 9 V, V_{ac} = 6 V

37. 1 V, it's lower

39. 20 V

41. 40 V (20), 100 V (50)

43. 12 V (4), 30 V (10), 15 V (5)

45. 20 V at *a*, 0 V at *b*, 10 V halfway

47. Change the 100 to 120

49. 5.5 V

51. R1 = 6 k, R2 = 1.5 k, R3 = 500

53. 19.2

55. Yes, 10 k

3 Parallel Resistive Circuits

1. 80 Ω

3. 14.3 Ω

5. 4 mA down

7. 120 mA (200), 480 mA (50), 600 mA (source)

9. 219.5 mA (82), 264.7 mA (68), 484.2 mA (source)

11. No. It is already the smallest current by an order of magnitude and this makes it smaller.

13. 333.3 µA (each 36 k), 250 µA (each 48 k)

15. 0.1915 mA (47 k), 1.765 mA (5.1 k), 5 mA (1.8 k)

17. 1.6 A (10), 0.4 A (40), 16 V

19. 50 mA (200), 25 mA (each 400), 10 V

21. 145.5 mA (3 k), 218.2 µA (2 k), 36.37 µA (12 k), 0.4364 V

23. 2.571 mA (each 25 k), 0.8571 mA (75 k), higher because there is one less path for current flow.

25. 5.714 mA (5 k), 2.857 mA (each 10 k), 28.57 mA (1 k), 28.57 V

27. The voltage is larger but now negative, so it becomes more negative.

29. 11.25 mA (2 k), 3.75 mA (6 k), 5 mA (4.5 k)

31. 1.714 k

33. 6.667 mA

35. 857 Ω

37. 3 k

4 Series-Parallel Resistive Circuits

1. 10 k with 30 k
3. None
5. 200 with 300
7. None
9. 12.5 k
11. 23.3 k
13. 106 Ω
15. 3 k
17. $V_a = 2$ V, $V_b = 1.34$ V, $V_{ab} = 0.662$ V
19. 0.4 mA (12 k), 0.4 mA (3 k), 0.8 mA (7.5 k)
21. $V_a = -18$ V, $V_b = -9.35$ V, $V_{ab} = -8.65$ V
23. 903 μA (3.3 k), 602 μA (10 k), 301 μA (20 k)
25. $V_a = 100$ V, $V_b = 59.5$ V, $V_{ab} = 40.5$ V
27. 15 mA (1 k), 3.75 mA (2.2 k), 3.75 mA (1.8 k)
29. $V_b = 2.87$ V, $V_c = 5.1$ V, $V_{cb} = 2.23$ V
31. 4 mA (2 k), 0.702 mA (5.1 k), 4.702 mA (source)
33. $V_a = 10$ V, $V_b = 5.263$ V
35. 0.2222 mA (30 k), 0.3333 mA (36 k), 0.8889 mA (source)
37. Yes, because they also have the same voltage.
39. $V_a = -22.4$ V, $V_b = -36$ V, $V_{ab} = 13.6$ V
41. 191.5 μA (47 k), 448.9 μA (5.1 k), 1.72 mA (3.9 k)
43. $V_b = 10.53$ V, $V_c = 7.643$ V, $V_d = 5.945$ V
45. 133.3 μA (left), 66.67 μA (middle), 33.33 μA (right)
47. $V_a = 22$ V, $V_b = 16$ V, $V_{ab} = 6$ V
49. 375 mA (100 k), 125 mA (60 k), 125 mA (240 k)
51. $V_a = 4.91$ V, $V_b = 3.27$ V, $V_{ab} = 1.64$ V
53. 50 mA (200), 25 mA (400), 25 mA (100)
55. $V_a = -1.2$ V, $V_b = -3.2$ V, $V_{ab} = 2$ V
57. 1.2 mA (9 k), 0.6 mA (82 k)
59. Yes, because they also have the same voltage.
61. $V_a = 3.54$ V, $V_b = 35.4$ V, $V_{ab} = -31.86$ V
63. 878 μA (1 k), 878 μA (2.2 k), 122 μA (18 k)
65. $V_a = -5.29$ V, $V_b = -3.31$ V, $V_{ab} = -1.98$ V
67. 4 mA (left), 2 mA (middle), 1 mA (right)
69. 133.3 Ω
71. 110 Ω
73. 2.55 k
75. 1 A
77. 4 mA

5 Analysis Theorems and Techniques

1. 4.26 mA in parallel with 4.7 k

3. 4.56 mA in parallel with 5.7 k

5. 26.4 V in series with 2.2 k

7. −40 V in series with 800 Ω

9. −18 V in series with 9 k

11. 2.55 mA left to right

13. 49.5 V

15. 8.18 V

17. 2.55 mA

19. 4.2 V

21. 1.25 mA

23. 0.7 V

25. 8.09 mA

27. 42.7 V

29. 0.696 mA

31. Yes, it's only current sources and resistors in parallel.

33. 1.17 mA

35. −15.65 V

37. −11.6 V

39. 4.03 mA (up)

41. 38.7 mV

43. 1.19 mA left to right

45. 7.2 V with 6.2 k

47. 2 mA with 3 k

49. 5.71 V with 1.71 k

51. 1.538 A with 13 Ω

53. 66.7 µA with 3 k

55. 1.92 mW. Yes, match R_{thev}.

57. 4.27 µW. Yes, match R_{thev}.

59. 3 k, 3 mW

61. 4 k, 4.44 µW

63. Replace 10 V/2 k with 5 mA||2 k, and 24 V/10 k with 2.4 mA||10 k

65. Replace 2 mA||20 k with 40 V + 20 k, and 0.4 mA||3 k with 1.2 V + 3 k

6 Mesh and Nodal Analysis

1. Loop ordering is left to right

    $9 = (1\text{ k} + 3\text{ k})I_1 - (3\text{ k})I_2$

    $-12 = -(3\text{ k})I_1 + (3\text{ k} + 2\text{ k})I_2$

3. 0.818 mA left to right

5. 3.14 V

7. $50 = (200 + 40)I_1 - (40)I_2$

    $-120 = -(40)I_1 + (40 + 75)I_2$

9. 1.03 A right to left

11. −2.3 V

13. First, combine the 4 k∥12 k = 3 k

    $-34 = (2\text{ k} + 10\text{ k})I_1 - (10\text{ k})I_2$

    $24 = -(10\text{ k})I_1 + (10\text{ k} + 3\text{ k})I_2$

15. 0.696 mA (up)

17. 12.27 V

19. $60 = (100 + 200)I_1 - (200)I_2 - (0)\ I_3 - (0)\ I_4$

    $0 = -(200)I_1 + (200 + 300 + 400)I_2 - (400)\ I_3 - (0)\ I_4$

    $0 = -(0)I_1 - (400)I_2 + (400 + 500 + 600)\ I_3 - (600)\ I_4$

    $-150 = -(0)I_1 - (0)I_2 - (600)\ I_3 + (600 + 700)\ I_4$

21. 47.7 mA right to left

23. −9.34 V

25. Loop 1 at top, then left to right.

    $30 = (15\text{ k} + 10\text{ k} + 9\text{ k} + 3\text{ k})I_1 - (3\text{ k})I_2 - (9\text{ k})\ I_3 - (10\text{ k})\ I_4$

    $25 = -(3\text{ k})I_1 + (2\text{ k} + 3\text{ k} + 4\text{ k})I_2 - (4\text{ k})\ I_3 - (0)\ I_4$

    $10 = -(9\text{ k})I_1 - (4\text{ k})I_2 + (4\text{ k} + 9\text{ k} + 5\text{ k})\ I_3 - (5\text{ k})\ I_4$

    $-10 = -(10\text{ k})I_1 - (0)I_2 - (5\text{ k})\ I_3 + (5\text{ k} + 10\text{ k} + 7\text{ k})\ I_4$

27. 0.86 mA left to right

29. −3.26 V

31. Loop ordering: left, top, bottom.

   $24 = (2\text{ k} + 8\text{ k})I_1 - (2\text{ k})I_2 - (8\text{ k})I_3$

   $0 = -(2\text{ k})I_1 + (2\text{ k} + 22\text{ k} + 800)I_2 - (22\text{ k})I_3$

   $0 = -(8\text{ k})I_1 - (22\text{ k})I_2 + (8\text{ k} + 22\text{ k} + 400)I_3$

33. 19.8 mA (down)

35. −22.4 V

37. $20 = (6.8\text{ k} + 10\text{ k})I_1 - (10\text{ k})I_2 - (0\text{ k})I_3$

   $0 = -(10\text{ k})I_1 + (10\text{ k} + 8.5\text{ k} + 30\text{ k})I_2 - (30\text{ k})I_3$

   $0 = -(0)I_1 - (30\text{ k})I_2 + (30\text{ k} + 20\text{ k} + 70\text{ k})I_3$

39. 0.34 mA left to right

41. 49.5 V

43. Current source converts to 10 V + 20 k.

   $8 = (5\text{ k} + 4\text{ k})I_1 - (4\text{ k})I_2$

   $-10 = -(4\text{ k})I_1 + (4\text{ k} + 20\text{ k})I_2$

45. 0.76 mA left to right

47. −2.73 V

49. $-6\text{ mA} = (1/24\text{ k} + 1/12\text{ k} + 1/9\text{ k})V_a - (1/9\text{ k})V_b$

   $-2\text{ mA} = -(1/9\text{ k})V_a + (1/9\text{ k} + 1/25\text{ k})V_b$

51. 4.03 mA (up)

53. −774 mV

55. Combine 15 k + 25 k = 40 k

   $2\text{ mA} = (1/20\text{ k} + 1/40\text{ k} + 1/5\text{ k})V_a - (1/5\text{ k})V_b$

   $0.4\text{ mA} = -(1/5\text{ k})V_a + (1/5\text{ k} + 1/3\text{ k})V_b$

57. 0.538 mA

59. 324 mV

61. $(300 - 20)\text{mA} = (1/40 + 1/10 + 1/100)V_a - (1/10)V_b - (1/40)V_c$

   $10\text{ mA} = -(1/10)V_a + (1/10 + 1/50 + 1/250)V_b - (1/250)V_c$

   $(-300 - 50)\text{mA} = -(1/40)V_a - (1/250)V_b + (1/250 + 1/25 + 1/40)V_c$

63. 177 mV

65. 5.94 V

67. This is deceptively simple. First combine the parallel 36 k resistors to a single 18 k. There is only one node of concern, node a. The 4 mA source enters this node. Define an exiting current, I_1, down through the 18 k. $I_1 = V_a/18$ k. Define an entering current from the voltage source, I_2. $I_2 = (12 - V_a)/30$ k. From KCL, 4 mA + $I_2 = I_1$. Substitute the current expressions into the KCL equation, expand and simplify. This results in a single node equation: 4.4 mA = (1/30 k + 1/18 k) V_a.

69. 1.25 mA left to right

71. 1.05 mA

73. 240 V

75. 720 V

77. 3750 V

79. 6 V

7 Inductors and Capacitors

1. 5 µF

3. 2 µF

5. 6 mH

7. 0.75 mH

9. 8 V across 10 µF, 4 V across 20 µF

11. 2 k gets 3.33 V, 5 µF gets 0 V,
    4 k and 3 mH get 6.67 V

13. 2 k gets 3.33 V, 3 mH gets 0 V,
    4 k and 5 µF get 6.67 V

15. 2 k gets 5.38 V, 5 µF gets 0 V,
    4 k and 3 k get 4.62 V

17. 2 k gets 5.38 V, 5 mH gets 0 V,
    4 k and 3 k get 4.62 V

19. 2 k, 10 k and 20 µF get 0 V,
    8 k and 12 mH get 20 V

21. 33 k gets 15 V, all other components get 0 V

23. 1.5 V

25. 0.2 A

27. 20 µs, 100 µs

29. 13.2 V, 24 V

31. 1 ms, 5 ms

33. 10 V, 15 ms

35. 3 ms, 15 ms

37. 20 ms, 100 ms

39. 0 V and 480 mA for both times.
    With 5 Ω, 2 V and 400 mA for both times.

41. 40 ns, 200 ns

43. 3 mA, −60 V

45. 833 ns, 4.17 µs

47. Replace 100 Ω with 2 k

49. 125 µF

Appendix C

Fun with Google Translate

Life is a matter of moments

La vie est une question de moments

Das Leben ist eine Frage der Zeit

La vida es una cuestión de tiempo

Жизнь это вопрос времени

Vita est causa temporis

人生は、時間の原因です

La vita è la causa del tempo

Life is the cause of time

Printed in Great Britain
by Amazon